The Evolutionary Origins of Life and Death

The Evolutionary Origins of Life and Death

Pierre M. Durand

The University of Chicago Press CHICAGO AND LONDON

The University of Chicago Press, Chicago 60637
The University of Chicago Press, Ltd., London

Published 2021
Printed in the United States of America

30 29 28 27 26 25 24 23 22 21   1 2 3 4 5

ISBN-13: 978-0-226-74762-0 (cloth)
ISBN-13: 978-0-226-74776-7 (paper)
ISBN-13: 978-0-226-74793-4 (e-book)
DOI: https://doi.org/10.7208/chicago/9780226747934.001.0001

Library of Congress Cataloging-in-Publication Data

Names: Durand, Pierre M. (Pierre Marcel), author.
Title: The evolutionary origins of life and death / Pierre M. Durand.
Description: Chicago : University of Chicago Press, 2021. | Includes bibliographical references and index.
Identifiers: LCCN 2020024774 | ISBN 9780226747620 (cloth) | ISBN 9780226747767 (paperback) | ISBN 9780226747934 (ebook)
Subjects: LCSH: Evolution (Biology). | Cytology. | Life—Origin. | Apoptosis. | Death (Biology). | Biological fitness.
Classification: LCC QH366.2 .D865 2020 | DDC 576.8—dc23
LC record available at https://lccn.loc.gov/2020024774

♾ This paper meets the requirements of ANSI/NISO Z39.48-1992 (Permanence of Paper).

Contents

Foreword

BY EUGENE V. KOONIN

"In this world nothing can be said to be certain, except death and taxes." The second part of this famous quip by Benjamin Franklin, obviously, applies only to us humans. The first part is more general, we know full well that all animals die, and so do plants, even if some, like redwoods, live incredibly long by our standards. However, this is where most people to some extent familiar with biology, and as far as I know, even most biologists draw the line: complex, multicellular organisms inevitably die but simple, unicellular ones, such as protists (unicellular eukaryotes) and bacteria, are thought to be, at least, in principle, immortal. Surely, the unicellular life forms perish massively, being killed by viruses, predators, starvation, and all kinds of adverse environmental factors. The common belief, though, is that they are not doomed to die in the same sense multicellular organisms are. In *The Evolutionary Origins of Life and Death,* Pierre Durand outlines a dramatic change to this perspective by synthesizing the discoveries in biology over the last two decades that compellingly show that most if not all unicellular life forms possess molecular mechanisms for programmed cell death (PCD). The existence and importance of PCD (often also called apoptosis, particularly in animals) in multicellular organisms is well known (in 2002, Sydney Brenner, H. Robert Horvitz, and John E. Sulston were awarded the Nobel Prize in Physiology or Medicine for the discovery of genes responsible for apoptosis in animals). Indeed, in animals and plants, PCD is essential for defense against pathogens, elimination of cells with impairments of the division cycle, and some developmental processes. Abrogation of PCD by somatic mutation leading to cell immortalization results in cancer. However, conventional thinking predicates PCD on multicellularity and separation of germline and soma: some somatic cells are sacrificed to benefit the organism as a whole. From this perspective, PCD in unicellular organisms appears puzzling, even paradoxical: what might be the benefit for a single cell organism to kill itself?

To explain PCD in unicellular life forms, a major shift in our understanding of evolution is required: group selection, i.e., selection that affects ensembles of individuals, must be recognized as a key evolutionary mechanism. If group selection is ubiquitous, so should be PCD: an ensemble of cells will benefit from some of its members sacrificing themselves to prevent the spread of infection or to provide nutrients at times of starvation. Should this be the case, one would have to conclude that there are no strictly unicellular life forms but rather a continuum of cooperation between cells, a striking realization in itself.

Group selection (also often referred to as *inclusive fitness*, a more technical term) has been for many years and remains a hotly debated subject in evolutionary biology. Some prominent theorists flatly deny the very existence of this phenomenon. And yet, the tide is turning, with multiple lines of empirical data and theoretical argument converging to show, beyond reasonable doubt, that group selection is one of the defining factors in the evolution of life. Indeed, in Durand's book, group selection emerges as a common denominator that unifies the origins of life and death, no less. The earliest steps in the evolution of life cannot be modeled or even reasonably imagined without selection affecting collectives of replicators, that is, without the emergence of group selection. Furthermore, as discussed at some length in the book, the evolution of life is punctuated by evolutionary transitions that involve changes in evolutionary individuality, or in other words, transitions to new levels of selection (origin of multicellular life forms is the most obvious case in point, and it occurred on multiple occasions in evolution). But, then, the same force that drove the evolution of life, from its inception, promoted the evolution of PCD, perhaps, concomitantly with the origin of the first cells, or at any rate, shortly after. At later stages of evolution, PCD itself promoted the evolution of complex organisms. The complete theory of group selection remains an unresolved task, but Durand presents a simple but powerful mathematical framework, based on the Price equation, to analyze this crucial evolutionary phenomenon.

I think *The Evolutionary Origins of Life and Death* will be an eye opener for many readers, biologists, and others interested in visiting the frontiers of today's conceptual thinking on evolution. This book, compact and simple but densely packed with information and ideas, presents the new edifice of evolutionary biology, a vibrant, progressing field that goes far beyond the confines of neo-Darwinism.

Preface

This book deals with the origins and evolution of the first "living" molecular systems and the first "dying" cells. I have purposefully placed the "living" and "dying" entities in quotation marks because their definitions are not as clear as we would like them to be. Furthermore, a detailed examination of how the two phenomena originated and evolved reveals that they are not always oppositional. They not only share some fundamental processes that facilitated their origins, but also exhibit coevolutionary features as more complex forms of life emerged.

With the advances and consolidation of research in the origins of life and cell death fields, as well as a general interest in these subjects, the time is right to organize these two areas into a single book. I hope that the subject material and integration of work and ideas from disparate disciplines will be a source of interest to a variety of researchers, academics, and scholars ranging from fundamental scientists in biology, physics, chemistry, evolution, and philosophy to students and perhaps even the lay public with a background and interest in natural history. Many of the recent advances (for example, the genomic and experimental systems) and the integration of this information into syntheses of life and death, have not yet found their way into general and evolutionary biology textbooks. I hope the syntheses presented at the end of each part will be useful for tertiary students. Most of the chapters should be accessible to those with a basic but sound knowledge of general biology and natural history. There are, however, a few sections and chapters that require much more detailed knowledge of cell and molecular biology or evolutionary theory (for example, chapters 6 and 13). For the non-specialists, these can be glossed over or even ignored without interrupting the flow of the book or impacting other chapters. At the same time, there may be some sections that more expert readers may find frustratingly simple. These are necessary to appeal to a more general science audience and I ask for your indulgence.

The section at the very beginning, "An introduction to the scientific study of life and death," will orientate the reader and provide a sense of the level of biological knowledge required to follow and, hopefully, enjoy the contents of this book. It also serves to highlight the motivation and reasons for writing this book. For non-academics or those without sufficient background in biology, the three main ideas in this book can still be appreciated by reading the three most important chapters. These are chapter 7 (a synthesis for the origin of life), chapter 14 (a synthesis for the origin of death), and chapter 16 (an explanation for how more complex life and death coevolved).

The reference list is extensive, and I have attempted to be as accurate and appropriate as possible. One of my gripes in science and philosophy, is that the most appropriate references are often not the ones cited. This happens all the time, especially by those at the early stages of their careers, when it is more convenient to attribute discoveries or advances to publications in high-profile journals or simply to those frequently cited by the research community (see also additional notes 3.2). I cannot claim to get this aspect right all the time—one needs a thorough knowledge of the history of science for that—but I have made every attempt to be as accurate and fair as possible. The references should be a helpful resource for those with a deeper interest, particularly students and academics who may wish to engage further with the subject material. I have to state upfront, however, that there is an enormous number of excellent publications in the origins of life and cell death fields. In many instances, I could easily have cited other contributions, but chose the ones I have because I am familiar with them. It is quite simply impossible to do justice to the many individuals who have made valuable contributions. I apologize for these omissions in advance. To offset some of this limitation there are further acknowledgments and references in the additional notes at the end of the book. I have also used these additional notes to make the reader aware of related issues or to mention alternate views that are in conflict with my own.

I was encouraged to think about this book during my time as a visiting scholar attending the program "Cooperation and the Evolution of Multicellularity" and conference "Cooperation and Major Evolutionary Transitions" at the Kavli Institute for Theoretical Physics, UC Santa Barbara, USA, in 2013. I am very grateful for my time there, which was supported in part by the National Science Foundation under grant no. NSF PHY-1125915. One is rarely afforded the privilege of immersing oneself completely in such an intellectually stimulating environment without the responsibilities of everyday university life. This indulgence allowed me to interact with others from diverse backgrounds and helped crystallize some of my own thoughts concerning the origins of life, death, and complexity. I am grateful to the organizers of that workshop (especially Greg Huber), and my interactions with Irene Chen, Eugene Koonin, Niles Lehman, Michael Lynch, David Queller, Joan Roughgarden, and many others were particularly encouraging. I thank Aurora Nedelcu and Armin Rashidi for many discussions concerning cell death, and Nisha Dhar, who conducted her PhD research on the origin of life in my research group. I pay special tribute to Rick Michod, with whom I spent an enriching postdoctoral fellowship in 2009/2010 at the University of Arizona, something that I valued greatly. I thank the editorial and publishing staff at the University of Chicago Press, especially Rachel Kelly Unger, Michaela Luckey, Scott Gast, and Christie Henry (who has since moved to Princeton University Press), for their enthusiasm and patience. I have received many helpful comments from other scientists and philosophers. I am particularly grateful for the discussions and feedback I have had from Troy Day, Sue Dykes (who also proofread the book), Eugene Koonin, Kevin Laland, Rick Michod, Aurora Nedelcu, Samir Okasha, David Penny, Grant Ramsey, Santosh Sathe, Sonia Sultan, Stuart Sym, and several anonymous reviewers. I appreciate the significant contributions of Nisha Dhar, Rick Michod, and Grant Ramsey to four of the chapters (please see my acknowledgments to them at the beginning of the relevant chapters). These chapters would have been much poorer without them. At the same time, any errors are mine alone. I am grateful to Eugene Koonin for the foreword and Viktor Radermacher for the figures in chapters 7 and 14 that capture my proposed syntheses for the origins of life and PCD. In addition to the NSF funding that covered my time at the Kavli Institute, I appreciate the funding and support received from Bruce Rubidge, director of the Centre of Excellence in Palaeosciences, and Marion Bamford, director of the Evolutionary Studies Institute, both at the University of the Witwatersrand. I am grateful for the support of friends and family, particularly Matthew McKay. And, of course, Donald.

PMD, 2020

Abbreviations

BMC	bacterial microcompartment
CVT	cytoplasm-to-vacuole targeting
DHA	dehydroascorbate
ET	evolutionary transitions
ETI	evolutionary transitions in individuality
FoL	forest of life
GNA	glycol nucleic acids
GWAS	genome-wide association studies
HGT	horizontal gene transfer
HRU	higher-level replicating unit
LRU	lower-level replicating unit
LUCA	last universal common ancestor
MGE	mobile genetic element
MLS	multilevel selection
PCD	programmed cell death
PNA	peptide nucleic acids
POD	programmed organismal death
PS	phosphatidylserine
ROS	reactive oxygen species
SCM	stochastic corrector model
SEM	scanning electron microscopy
TA	toxin-antitoxin
TE	transposable element
TEM	transmission electron microscopy
TNA	threose nucleic acids
ToL	tree of life
TUNEL	transferase dUTP nick-end labeling

An introduction to the scientific study of life and death

The origin of life and its inevitable association with death is one of the most intriguing and enigmatic areas in all of biology. There has always been an abiding interest among scientists, academics, and the lay public in the driving forces and mechanistic processes by which life arose on Earth. What constitutes the beginning of life is still very much debated, but what is clear is that molecular life, or at least a collection of compartmentalized molecules that together encompass a program for the properties that are associated with life, existed long before cells as we know them today emerged. Once a system with the features that we associate with life existed, it would have been possible for such a system to die. Death may have occurred in many ways, but at this early stage death was almost surely the result of aged and damaged molecules or accident. What is surprising from an evolutionary perspective is that a new way of dying evolved. One that is active and non-incidental. One supported by a genetic program. The characteristics of the very first forms of programmed death are unknown. But what is clear is that from the time that single-celled life existed (and certainly long before multicellular life arose), there was a programmed form of cell death. This book deals with the evolution of this form of programmed death in the world of single-celled organisms.

An investigation of the origin and evolution of life focuses on the selective pressures that promote life—the selection of traits that stabilized the first living systems and allowed them to evolve. But a study of the evolution of death (in this book the reference to death is in the context of unicellular organism unless specifically indicated otherwise) seems, at first glance, a peculiar endeavor. When a genetic program for death in unicellular organisms was uncovered in the latter half of the twentieth century it puzzled evolutionists. Why would unicellular organisms harbor a suicide-like genetic program, for surely natural selection would not favor such an obviously lethal trait? The phenomenon of what became known as programmed cell death (PCD) was first discovered in multicellular organisms, where it is part of organism ontogeny and tissue homeostasis. But its observation in unicellular organisms raised all sorts of intriguing questions. Despite this counterintuitive finding (since in unicellular organisms death of the cell equates to death of the organism), many of the genetic and protein elements of the PCD molecular machinery have now been identified in diverse unicells and an argument can be made that PCD is almost as old as cellular life itself.

Why and how two of the most foundational biological innovations (life and cell death) emerged, and how they are connected evolutionarily and mechanistically, are the subjects of this book. The *why* question deals with ultimate causality: the explanation of origin and evolution in terms of selective pressures, adaptation, and non-adaptive events. The *how* question deals with proximate causality: the mechanisms by which the processes occurred as revealed by biochemistry, molecular and cell biology, and related disciplines. To students and non-experts, it is worth cautioning that the two questions can easily be confused, and many evolutionists will be quick to point out this error.

Life and death are philosophically and conceptually connected; the one exists in the absence of the other. The two terms are typically used to describe two mutually exclusive states in biology. Thus, a living entity, whether a single organism or a complex biological ecosystem, can be alive only if it is not dead and vice versa. This was certainly the case at the origin of life and death. However, as the subject is explored further, and as more complex life (like groups of interacting cells) emerged, it becomes clear that life and death processes are not always oppositional. They exhibit some coevolutionary features. In multicellular organisms, cellular death may, for instance, serve to promote the life of the organism of which they are a part.

This book is structured around three main ideas. Part 1 deals with our understanding of the first living molecular systems and proposes a synthesis that integrates the evolutionary, theoretical, computational, and biochemical knowledge of the origin and evolution of life. Part 2 investigates the origin and evolution of PCD and culminates in a synthesis of how a genetic program for death may have evolved in unicellular organisms. The third main "big idea" for this book is in part 3, which reveals how cellular life and PCD coevolved en route to more complex life forms. One of my aims has been to integrate much of the current information from disparate disciplines. These include philosophy of biology, theoretical biology, molecular biology and biochemistry, cell biology, genomics and computational biology, and of course evolutionary theory, which is the central organizing framework around which all of biology is structured.

Part 1: The origin of life

The field of biology has reached a point where a broad explanation for the origin of life is within our grasp. Of course, significant gaps remain and seeking a definitive explanation is likely never going to be realistic; but a crude step-by-step account of how it could have happened is certainly possible. Part 1 begins by examining practical philosophical and conceptual considerations of what life is (chapter 1) and the fundamental differences between the abiotic and biotic worlds (chapter 2). What is the quintessential feature of life and how does one capture it philosophically, theoretically, and empirically? The theoretical underpinnings for life's origin are covered in chapter 3 and include the theory of hypercycles and quasispecies, group selection models, contributions from the fields of genomics, phylogenetics, and bioinformatics, as well as the enduring concept of Gánti's chemoton. Chapter 4 introduces the chemistry of the first biologically relevant molecules starting from the generation of nucleobases, sugars, and nucleotides through to the passive chemical replication of small RNA molecules. The emergence and enzymatic replication of the first replicators is covered in chapter 5. The empirical evidence and the explanatory gaps for the emergence of complex replicating molecules are highlighted. Chapter 6 discusses the origin of life as an evolutionary transition in individuality. The concepts of the levels and units of selection as well as issues concerning group selection are introduced. These are discussed again in part 2 (especially chapter 13) and given a more informal treatment in the additional notes. Chapter 6 explains how groups of replicating networks became functionally integrated and formed the very first proto-genomes. The basic features of all of life are traced back to these original collections of molecules. The first six chapters lay the foundation for the final chapter in part 1 (chapter 7), which is the first main aim of this book. It integrates the disconnected information from different disciplines and proposes a synthesis for the origin of life.

Part 2: The origin of death

Before the turn of the last century there was the widespread belief that single-celled organisms were immortal and that death in unicellular life was secondary to external factors like nutrient depletion, physico-chemical damage, predation, or the accumulation of metabolic waste products with subsequent loss of viability. A genetically encoded obligatory death program like PCD was considered a hallmark of multicellular life. However, in the latter decades of the last century, a PCD-like phenotype was observed in diverse unicellular organisms. What started as a minor curiosity become a fundamental problem in biology. The existence of death in unicells was described as "a beautiful evolutionary problem" (anonymous National Institutes of Health grant reviewer, 2013), "enigmatic," "counterintuitive," "confusing," or simply an "anomaly." Part 2 focuses on this evolutionary conundrum. Chapter 8 introduces the phenomenon of death and considers the applied philosophy and terminology of different forms of death. Chapter 9 illustrates the phenotypes of death in unicellular organisms. The ultrastructural changes are used as a tool to explain the processes by which unicellular organisms die in a programmed way. A range of organisms are used as examples, including prokaryotes and photosynthetic and non-photosynthetic eukaryotes.

Chapter 10 details the mechanisms of PCD as well as the measures of PCD in different microbes. All the major crown groups of unicellular organisms contain elements of PCD and the evolutionary relationships between lineages are used to explain the ancient origins of death. The experimental and theoretical evidence for PCD as an adaptation is discussed in chapter 11. The arguments describing PCD in unicells as a non-adaptive process are provided in chapter 12. Chapter 13 introduces the issue of multilevel selection theory, which is a central issue if we wish to understand and capture the concepts of PCD in unicells. Has PCD been selected for, and if so how? Chapter 14 concludes part 2 with the second main aim of this book, which is a synthesis for the origin and maintenance of PCD in the unicellular world. The reader is taken through the putative steps from the ancestral state of immortality or death by accidental means, to the current observations that inform our understanding of what PCD in unicellular organisms fundamentally is.

Part 3: Origins of life and death, and their coevolution

The syntheses of the origins of life and death reveal that they are not always oppositional. They also share some evolutionary processes like evolution by group selection, which has been one of the most contentious debates in all biology. Despite its controversial history, it seems that group selection processes were essential for the evolution of the two most fundamental events in all of biology. Chapter 15 highlights the essential and unexpected role that group selection played and places this in the context of the history of group selection theory. Chapter 16 is the third main motivation for this book and examines the coevolution of PCD and more complex life. It has become clear that in some circumstances PCD is essential for sustaining life. In other instances, it appears that PCD has played a role in the emergence of complex microbial communities, sociality in the unicellular world, and multicellularity.

PART ONE

The origin of life

The only thing I know for certain about how life began, is that it did.

1
Philosophical considerations and the origin of life

Attempts to understand the origin of life and the history of living systems have had a profound impact on human thought and civilization. The question of life's origin has possibly been a part of human curiosity from the very beginning, from the time that consciousness in our primate ancestors had evolved to a point where they could pose such conceptual questions. It is an endeavor that has intrigued seasoned philosophers, scientists, and other academics as well as the lay public in equal measure. Before and during the Age of Enlightenment and into the modern scientific era, an explanation for life was proffered by thinkers of the time. Every ancient culture and civilization to the more recent monotheistic Judeo-Christian-Islamic traditions had, or still has, their own version of events (fig. 1). In the Western tradition of philosophy, the ancient Greeks (although it should be remembered that philosophy pre-dates ancient Greek civilization [for example, van de Mieroop 2015]) attempted explanations based on reason even without any of the mechanistic knowledge that we have today. In the pre-Socratic era, Thales the Milesian (639–544 BC) and his student Anaximander of Miletus separated rational thought and observation from theology with the aim of explaining natural phenomena in a reasoned way (see chapter 6 in Finley 1963). Scientific methodologies brought about an empirical examination of life's origins. While it is unlikely that the puzzle of life's origins will ever be answered definitively, the field has reached a point where a reasonable explanation, albeit crude and with many gaps, is possible. The hypotheses are testable and rooted in our knowledge of biochemistry, geochemistry, physics, mathematics and theoretical simulations, molecular and evolutionary biology, and philosophy. The aim of this book is to integrate much of what is known about the origin of life from these diverse disciplines and to present and discuss one of the likeliest ways in which life could have arisen. The final chapter of part 1 (chapter 7) is a synthesis of the ideas and empirical studies dealt

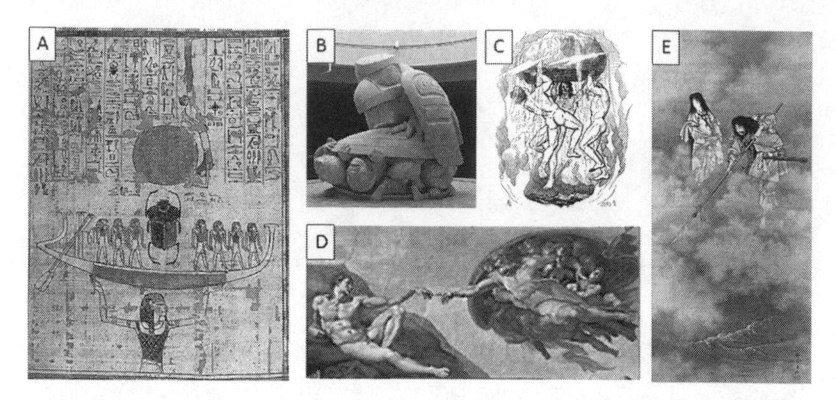

FIG. 1. Cultural accounts of life's origin. Creation myths have had a profound impact on human civilization and cultural evolution. All known, past and present, civilizations proffered some or other account of how life (humans and all living things) began. The creation myths are sometimes focused on the universe and all life contained in it (example A). In other instances (example B), the creation myth focuses only on humans and animals. The myths appeal to the idea that before life as we know it existed, there were supernatural entities (a god or gods, or animals) that created the universe of life. Five creation myths are illustrated. (A) Nun, god of the waters of chaos, lifts the barque of the sun god Ra (represented by both the scarab and the sun disk) into the sky at the beginning of time (artwork created c. 1050 BC); (B) The Haida people's creation story explains how the Raven opened a giant clam from which humans emerged (sculpture by Bill Reid [1920–1988] in the University of British Columbia Museum of Anthropology); (C) In Norse mythology Odin and his brothers Vili and Ve slew the god Ymir and created the universe and everything in it from his corpse (illustration by Lorenz Frølich [1820–1908]); (D) Artwork by Michelangelo (1475–1564) in the Sistene Chapel illustrates the Judeo-Christian-Islamic myth of Adam's creation by God; and (E) Shinto gods Izanami and Izanagi created islands from water droplets. Living on the islands Izanami produced many offspring, which included more islands, deities, and humans (painting by Kobayashi Eitaku [1843–1890]).

with in the first six chapters. But before exploring the theoretical and experimental advances it is worthwhile considering what is meant by life and living systems.

What is life and is it important to have a definition?

Szostak argues that "attempts to define life do not help to understand the origin of life" (Szostak 2012). It is probably true that a definition per se will not help us understand its origins; however, having some idea of what is being referred to when speaking about the properties and constituents of the earliest forms of molecular or cellular life is still helpful. Even if defining life does not help with a mechanistic understanding of life's origins, definitions of what we mean by a living system will at least help operationalize some of the questions (Bich and Green 2018) and contribute to a philosophical and conceptual appreciation of the transition from non-living to living systems (Knuuttila and Loettgers 2017; Mariscal and Fleming 2017). At the same time, while a consideration of what is meant by life is helpful it should not (as all philosophy should not) stray too far from common sense—a criticism that is sometimes (perhaps unfairly) aimed at theorists and philosophers.

The abiotic-biotic boundary is discussed in detail in chapter 2. For the moment, an appreciation of what is meant by the "origin of life" is necessary because it means different things to different people (see the editorial introductions and references in the special issues, Walker, Packard, and Cody 2017, and Gayon et al. 2010). There are numerous definitions and concepts of what, if anything, distinguishes the living from the non-living world (table 1 in Prud'homme-Généreux 2013). Different disciplines have their own perspectives (Jeuken 1975; Mariscal et al. 2019), and debates aimed at answering the question "what is life?" usually end in a stalemate. None of the concepts of what life is satisfies everyone (Popa 2004), which as we shall see reflects the consensus that life's origin cannot be captured by a substance or event. The abiotic-biotic transition was a process, not a moment in time, and the biomolecules that gave rise to life were chemically dynamic. Basic or pure (as opposed to practical or technical) philosophers are interested in the very notion of what life fundamentally is, its nature and what it means (Farley 1986; Kolb 2016; Ruse 1997). This is usually quite distinct from the interests of biologists (although the distinction itself is a philosophical question), which includes evolutionary biology and the mechanistic disciplines, each of which focuses on what it considers to be the most essential element of life. Astrophysicists, geophysicists, and biochemists are acutely aware of the importance of chemical and electromagnetic energy in the early reactions and may, for example, claim that life began by superflares (Airapetian et al. 2016). Of course, what is meant is that the energy required for life-giving chemical reactions may have come from solar radiation. On their own these were just physical or chemical phenomena. Protein biochemists may claim that life began by the formation of amino acids (Carter and Wolfenden 2015), but again amino acids or short peptides are not living although they form part of life and may have been essential for life to begin. To a cell biologist the organizational structure is the important component and life's origin may become interesting to them only once the earliest cells existed, concluding that archaea-like organisms were the first forms of life. But of course, others will want to know where the cells came from. In the absence of cells, a geneticist would want to see some form of replicating hereditary material and consider this material living regardless of whether it existed in cell-like compartments or not. An evolutionist is focused on processes like natural selection, which certainly captures a fundamental feature of living systems. But the purely evolution-based conceptual definitions can run into problems when more mechanistic questions are asked about the kinds of molecules present at the very beginning of life. Disparate fields agree to disagree and usually settle on a list of criteria or components that define living systems.

The question can be raised: is it even helpful thinking about the philosophical and biological questions of what life is and how it arose? (Szostak 2012). I think so, yes, even if it is much easier to pose the questions than provide the answers (Durand and Michod 2011). Not having satisfactory answers, does not mean that important insights will not emerge from the discussion. Even when considering the most fundamental concepts in biology, like the "species concept," there is no definition that is universally agreed upon (Hey 2006). There are still new definitions emerging to determine what constitutes a species (Thackeray and Dykes 2016; Thackeray and Schrein 2017; Stat et al. 2012). In the end, researchers often select a definition that is most appropriate for the questions they pose. That is the nature of biology. Boundaries are seldom discrete and exceptions to the rule are common. One can argue that the absence of a definitive species concept has actually aided speciation studies by highlighting the important elements. As with all living things, speciation is a process, not a discrete event (Thackeray and Schrein 2017). And so it is with life's origins (and the origin of death—see part 2). Even if there is no agreeable solution to the question "what is life?" thinking about the answer encourages an appreciation of the important components, processes, and features and helps formulate testable hypotheses (Bich and Green 2018).

Broadly speaking, two groups of origin of life biologists have emerged (the purely philosophical approaches are not considered here; see chapter 2): those focusing on "metabolism first" and those insisting on "replication first" (Lazcano 2010). The debate between proponents of metabolism and those of replication continues unabated with both approaches subject to criticism (Pross 2004). Understandably, replication first is favored by evolutionists since there can be no evolution without reproduction. However, the replication first proponents are not without their critics, particularly those who claim that without taking into account early metabolism, theoretical and laboratory studies are rendered meaningless (Shapiro 2000). Replication requires energy and nucleotide monomers, which could only have emerged through metabolic reactions. Without including these metabolic considerations, there is no way of knowing many of the biochemical features of replicators and modeling their dynamics can be unhelpful if the biochemical constraints are not taken into account. The questions concerning replication are discussed in more detail later, but it is worth considering briefly the metabolism first scenario to highlight some of the important biochemistry.

The case for metabolism first focuses on the generation of energy-rich organic compounds and information molecules that were essential for life. The argument is that metabolism must have existed prior to replicating capabilities since the information molecules like ribonucleotides as well as the energy compounds required to polymerize them could have emerged only via metabolic processes. The proposition is that metabolic reactions emerged spontaneously and at some subsequent stage a nucleotide-based replicative capability was incorporated into the existing system (Shapiro 2007). Metabolic reactions like the formose reaction (first described by Alexander Butlerov) (Leicester 1940) or a reverse Krebs cycle (the Krebs cycle is a source of chemical energy for molecular biosynthesis and polymerization) can be driven by mineral catalysts (Orgel 2000; Wachtershauser 1990), although critics claim the likelihood of this happening was remote (Orgel 2008). The iron-sulfur world hypothesis proposes that the first cells were enclosed in microscopic metal casings, which served as catalysts for biochemical reactions and at the same time concentrated the reactants and products in space and time. There is some evidence to support this; the discovery of iron-sulfide bubbles in the deep ocean provides a possible environment where such a system could have occurred (Miller and Bada 1988; Russell and Hall 1997). It is suggested that the spontaneous metabolic synthesis of lipids in such bubbles permitted the escape of lipid-enclosed cells from their metal origin and eventually gave rise to the earliest prokaryotes. Supporting this hypothesis is the observation that bacteria and archaea have membranes composed of different types of lipids, suggesting that they emerged independently from compartments like the proposed metal cells (see fig. 1 in Koonin and Martin 2005). Furthermore, the thermophilic archaea and bacteria that thrive on sulfur, methane, and hydrogen are at the root of the evolutionary tree and are commonly found near the hydrothermal vents that mimic the conditions proposed by the iron-sulfur hypothesis.

The value of the metabolism first hypothesis lies in the fact that it draws attention to the important chemical processes like the spontaneous generation of energy-rich molecules, which had little to do with the replication process itself but were essential for life to begin. So, excluding the metabolism first component limits the replication first explanations. Modes of replication that do not take into account the chemical environment are often unrealistic and investigating them theoretically is sometimes a technical pursuit without bringing us any closer to what actually happened. At the same time, without replication there is no population on which evolution can act. In order for evolution to occur, a population of replicating individuals (for further reading on individuality see Buss 1987; Goodnight 2013b; Santelices 1999) is required, and excluding heritability and replication makes no sense when considering the origin and evolution of life (additional notes 1.1). The truth is that while researchers may focus on either metabolism or replication, the two processes coevolved.

Returning to the issue of a definition of life. To assist with generating testable hypotheses, the most helpful definitions have often been those that are specific to a particular question. For example, the working definition settled upon by one group of origin-of-life scientists appears customized to the questions they ask. To understand the origin of life, they state that life is "a self-sustaining chemical system that can evolve by Darwinian evolution" (see the foreword in Joyce 1994), which is helpful insofar as one wants to understand life's origin in these terms. "Darwinian evolution" refers to natural selection and definitions like these are clearly useful. Natural selection is one of the organizing principles in biology (for the original publications by Darwin and Wallace see Darwin 1859, Darwin and Wallace 1858) and the above definition allows one to operationalize the problem and develop testable hypotheses. But definitions like these can also be limiting because they explicitly exclude evolution by means other than natural selection (see additional notes 1.2), which also played a role (Denny and Gaines 2002; Koonin 2011; Lynch 2007; Ramsey and Pence 2016). The above definition is tailor-made for the model system employed in this instance, and studying the origin of life in the terms employed in the definition will always answer only part of the question (the logic of definitions and terminology and how they impact interpretations of causality in evolution is dealt with in greater details in chapter 6 and additional notes 6.4). Furthermore, one can never be sure whether developing a chemical system in the laboratory that fits with the definition is more a technical feat than a true representation of what happened. Perhaps this was one of the reasons why Szostak made the point that attempts to define life will not help understand its origin.

At the other end of the scale, definitions may be accurate but at the same time too general and not very helpful to experimentalists. The statement, for example, that something is living if it has a life history must be true and is useful in the sense that one can measure life history traits; but the logic is circular. I am partial to the argument that life is an emergent phenomenon (see additional notes 1.3). This abstraction is, as discussed later (chapter 2), inclusive of many of the properties associated with life, but admittedly it is not a definition that is particularly helpful should one wish to operationalize the quantifiable aspects of "emergence." It may capture a quintessential component of life but gives no idea of how to investigate the problem (but see Hogeweg and Takeuchi 2003; Takeuchi and Hogeweg 2009). Furthermore, while life is an emergent property, not all emergent properties are living. Nevertheless, the property of emergence is certainly one way of capturing life's origin and features in the synthesis presented in chapter 7.

How, then, can we avoid being paralyzed by the limitations and inconsistencies when thinking about what life is? A useful approach is to accept that, as with almost everything else in biology, the origin of life was a process. Pluralist approaches and integrating information from different perspectives are necessary to appreciate the process fully (see for example, Dieckmann and Doebeli 2005; Gontier 2015). To study it, one needs to appreciate the important properties and minimum component parts that make the system functional. None of these on their own is sufficient, but each is necessary. Penny provides a useful framework for interpreting the concept of life and asking questions about how it began (Penny 2005). He makes the point that it is easier to understand life in terms of properties rather than definitions, the important elements being an energy source (and energy gradient), basic biochemical reactions driven by the energy gradients, organization (membranes, compartmentalization, and separation from the external environment), and self-reproducibility (genetic heritability, information transfer, evolvability). The process that defines life comprises these components.

For the remainder of part 1 it is important to provide some context. All the contributions concerning the philosophical, biochemical, physical, and molecular biology nature of life's origin are important, but the evolutionary questions are what percolate through the chapters that follow. From a mechanistic point of view, all the component parts that are mentioned as being a part of living systems, are essential. There is strong evidence that metabolism, replication, energy gradients, compartmentalization of biomolecules, and structures (cellular or otherwise) are all part of the process and essential to understand the mechanisms by which life emerged. The quintessential feature, however, is the *process* that gave rise to life (the emergent property) rather than discrete events or whether one type of substance was more important than another. At the same time, the evolutionary process must be rooted in what we know about the early chemistry. This is an important consideration, since without relevance in the material world, the theoretical and evolutionary studies can become mathematical and conceptual feats rather than practical solutions, sometimes leading to situations like Zeno's paradoxes. In addition, one must also consider the most parsimonious explanation, which is why the RNA world hypothesis, developed initially by Carl Woese, Francis Crick, Leslie Orgel, and others (for recent reviews see Higgs and Lehman 2015; Robertson and Joyce 2012) and which has been adopted and expanded by many other researchers, is the model system discussed in most detail (chapters 4–6, see also additional notes 1.4).

2
The biotic world

What are the fundamental differences between the biotic and abiotic worlds? Detailing some of these differences is important for the interpretations of the origins and evolution of life and death—the back and forth transitions between the first living and non-living systems. At face value, the difference seems obvious. Living material has hereditary properties, grows, and reproduces, while non-living substances do not. I have heard some non-life scientists argue that salt crystals do grow and produce more copies of the crystalline structure, but there is no meaningful hereditary component or active replication process that allows them to evolve by natural selection (additional notes 2.1). These fundamental differences between the living and non-living worlds have been widely discussed in natural philosophy (Deplazes and Huppenbauer 2009; Huneman 2015). At the same time, however, both living and non-living materials are reducible to atoms and molecules and the interactions between them. Life is made of the same matter as non-living material and both are subject to the same chemistry (the laws of thermodynamics, etc.) and physics (the laws of motion, quantum mechanics, etc.). Due to the living and non-living worlds being composed of the same materials, it is frequently argued that there are no fundamental differences between life and non-life. They are both reducible to the same sorts of things even though some of their properties may differ. In my opinion, however, the more important question one is compelled to ask is this: are there additional laws that apply to systems and phenomena in the biotic world that do not have any relevance in the abiotic one? In other words, are there properties or phenomena that occur in the living world that are not adequately captured by the laws of chemistry and physics?

Methodological reductionism sensu stricto presents a framework in which all material (regardless of whether it is living or not), its properties and associated phenomena, can ultimately be explained by the simplest component parts, reducing the entity of interest in a stepwise fashion and providing explanations for the simpler components (Fang and Casadevall 2011). The entity of interest is reduced in a stepwise fashion until the simplest parts are reached. Phenomena attributed to the simplest parts are used to explain those of the more complex entity. The claim is that if there is a thorough understanding of the atomic components and the interactions between them, the information will also explain the properties of their collectives ("atomic" is used here in the etymological sense and refers to the irreducible components, whatever they may be, that cannot be dissected any further). This approach has been successful in the physical, chemical, and biological sciences. However, many philosophers and scientists argue that while there is no doubting the success of this approach, there are some properties, particularly in living systems, that defy this methodology (Dupré 2010). In some instances, it comes down to a question of the language used by the researchers, but there are strong arguments to support the notion of irreducibility. Later in this chapter, these arguments will lead to a more significant claim that there are indeed, rules, laws, or principles (see additional notes 2.2 for terminology) that capture aspects of the living world and which have no meaning in the purely physico-chemical one. There are biological principles that explain the phenomena we associate with life and that are not meaningfully explained by the fundamental laws of physics and chemistry. At the outset, it is important to state that the claim in this book is not that there is anything *super*natural about life, although there is a vast body of work in pure philosophy that argues for theological interpretations of life (for example, Platinga 1978). In other words, my claim is not that life falls outside the laws of nature. Life is made up of the same matter that makes up non-living things and is subject to the same laws of physics and chemistry (Oparin 1953, see also additional notes 2.3). Any principles in biology cannot override or contradict the laws of physics and chemistry. For scientists there is no confusion. The point is that the reductionist approach does have limitations, and these are noteworthy when studying biology and particularly relevant when trying to distinguish biota from abiota (for opposing views see Fox Keller 2010; Dupré 2010). Emerging out of the discussion in this chapter is the claim that there are phenomena and properties of living systems (especially when examining them from the point of view of evolutionary biology) that are not adequately captured and described by the laws of physics and chemistry (Wolf, Katsnelson, and Koonin 2018).

At some stage in the future, the formulations describing the physical world may be extended to include biological phenomena. In fact, some researchers explicitly attempt to formulate the "physical principles of biological evolution" (Katsnelson, Wolf, and Koonin 2017), but as things stand, our understanding of biology is not sufficiently captured by physics and chemistry. It is propitious to provide examples of the properties and phenomena that support this argument since the über-reductionists sometimes dismiss this as fanciful or outlandish. There are two aspects that are central to the claim that there are additional biological principles necessary to capture the living world adequately. The first is the notion of emergent properties and their role in understanding the evolution of complexity. Emergent properties themselves are not unique to biology but they are particularly relevant to life because the properties that emerge in living systems lead to processes and phenomena that do not occur in abiotic systems. The second is the notion of functionalism, that a meaningful interpretation of the phenomena that arise from life's emergent properties is not achievable by applying the physical and chemical laws alone.

Emergent properties and living systems

As the word suggests, emergent properties emerge either from the interactions between the component parts or as a result of the number or types of components (Humphreys 1994). The property of interest exists only because of the number or kinds of elements and the relationships between them in the system. Separating out the component parts and manipulating them experimentally to investigate them in isolation may destroy the property of interest. One of the commonly used examples taught routinely in basic chemistry classes to explain the concept of emergence pertains to non-living material. Hydrogen and oxygen are covalently bound to form a water molecule, which exhibits polarity. Molecular hydrogen (H_2) and oxygen (O_2) are non-polar, so reducing water to its component parts eliminates polarity, which is key to explain the properties associated with water. At the same time, a single water molecule in isolation does not exhibit the physical properties of collections of water molecules. To study the flow dynamics of water, for example, requires many water molecules. This is not only essential to study them experimentally (isolating one water molecule is a challenge), but the interactions between the molecules are what impart the property of interest—in this case flow. The two properties discussed here (polarity and flow) demonstrate the kinds of phenomena that arise from collectives. In the first example, polarity is the property that emerges in the new entity formed by combining two components (hydrogen and oxygen). Polarity is emergent. In the second example, the flow properties emerge because of the nature of the aggregate. The property is aggregative. This distinction between properties that result from emergence and aggregation (Thompson 2000) will be noted in some of the biological examples used later.

Emergent properties are necessary for living systems to evolve. When genes were first discovered as the basic functional unit of DNA it was assumed that the flow of information from the gene to RNA and protein would neatly account for all traits or phenotypes. It quickly became apparent that the genotype-phenotype correlation was not 1 to 1 and that in most instances there was not a simple causal chain of molecular events from the genetic code to the phenotypic trait (Hull 1974). Complex traits can be traced back to genes or genetic regions, but the connection is neither absolute nor definitive. Of course, there are monogenic traits. The sickle cell disease phenotype in humans, for example, is explained by a single genetic disorder and the causal chain of events between genotype and phenotype is clear. Reductionism works well in cases like this, but for most phenotypes that is not the case, and Heng's genome-centric approach is in most instances more appropriate than the gene-centric one (Heng 2009), especially now that it is known that so many phenotypes are integrated at a genetic level (Boyle, Li, and Pritchard 2017). Furthermore, the more complex the trait the more convoluted the genotype-phenotype relationship. In fact, in many cases the causal relationship between the genotype and phenotype is not known and only associations are possible. An enormous number of GWAS (genome-wide association studies) projects have re-enforced this. GWAS have been used to investigate an array of organism traits and have become the order-of-the-day methodology to investigate the genetics of complex human diseases. In most cases, there is an association between the phenotype and chromosomal regions, single-nucleotide polymorphisms, and allelic and copy number variants. Usually, the findings are statistical correlations and the genotypes associated with the trait are neither necessary nor sufficient.

How should one approach the genetic basis of complex traits? The answer lies in our understanding that the genotype is relevant only because of the context in which it finds itself. The co-occurrence of other genetic regions, as well as epigenetic regulation, environmental influences, and the dynamic interactions between protein molecules, all contribute to the phenotype. As discussed earlier using the examples of polarity and flow, the phenotype can be an emergent or aggregative property. It is the result of the collective and the interactions between its components and cannot be quantified purely in terms of its component parts (but see additional notes 2.4). There is nothing mysterious about emergent properties, although the philosophical analyses of what they fundamentally are and their significance in the natural word are compelling (see the example of consciousness below). They are natural phenomena that occur as a result of the interactions between biomolecules, which are subject to the laws of physics and chemistry (for a philosophical review of emergence see Vesterby 2011). A clear demonstration of the relationship between the physico-chemical properties of matter and the emergent properties of life was achieved by freezing brine shrimp at temperatures near absolute zero and resuscitating them by slow thawing (Skoultchi and Morowitz 1964). This experiment elegantly illustrated that all the information for life is stored in the configuration of matter—atoms, molecules, and interactions between them. However, the properties we associate with a living brine shrimp (movement, respiration, metabolism, reproduction, etc.) are the result of the collective and the interactions between components. The properties could not be reduced to individual atomic, molecular, or cellular structures. The issue is that emergent properties in living systems are often inaccessible to the current reductionist approach. Life is a natural property of matter. It is not exempt from the laws of physics and chemistry, but to fully appreciate it one needs additional tenets.

Functionalism in biology

Related to emergence is the notion of functionalism, which distinguishes living and non-living. (Autonomy and organization are sometimes included in definitions of life and functionalism, but this is not explored here. For further reading see Ruiz-Mirazo and Moreno 2012; Moreno Bergareche and Mossio 2015.) In biology, functionalism refers broadly to the idea that what makes something living is the way it functions, where the function is derived from a mechanism and is governed largely by natural selection (Huneman 2013). Examples of biological functionalism are reproduction, metabolism, behavior, and consciousness. Brute matter exhibits no such functionality. Functions that separate biota from abiotic matter may include any phenomenon or trait that evolves, from molecular pathways to cellular activities and organismal properties like respiration, digestion, etc. At the origin of life, the hypercycles and collections of ribozymes discussed in the next chapter are good examples of biological functionalism. In some respects, functions may appear to have purpose (i.e., to be goal-directed), or at least that is how they were sometimes described. For example, the statement "the function of the heart is to pump blood" implies that the end-goal of pumping blood was why the heart emerged. Teleological explanations like these use future, goal-directed processes to explain present traits. Prior to the major advances in evolutionary theory, the early philosophers like Plato, Aristotle, and others frequently used teleological arguments to explore the functionality that distinguished life from non-life (Ariew 2002). Following the progress made by Lamarck, Darwin, Wallace, and others, evolutionists (Mayr 1974) and philosophers (Kant 1996) rightfully found this language problematic, primarily because it implies backwards causation. Biological functionalism, which is sometimes found at the core of discussions about what is living, has, subsequently, become part of a much larger philosophical debate on functionality, like the philosophy of mind (Wouters 2005). Perhaps the most wholesome example of biological functionality is consciousness and the mind-body problem. This example is worth further consideration because it touches on all the aspects related to functionalism that are frequently used in debates concerning the abiotic-biotic divide.

Much of the thinking around the mind-body problem traces back to Descartes and his critics (Descartes 1642) (for a literal translation see Heffernan 1990). In its simplest form, the mind-body problem concerns questions about what mental and physical states are, and what the relationships are between them. At the center of the debate lies the phenomenon of consciousness. As Nagel states: "Consciousness is what makes the mind-body problem really intractable" (Nagel 1974). Consciousness is an example of an emergent property in the living world (as opposed to the example of water molecules in the physico-chemical world, discussed above) and at some level or another includes most, if not all, aspects of the abiotic-biotic divide. Related to the phenomenon of emergence, consciousness also exhibits a new level of complexity that can be selected for by natural selection (natural selection is itself a feature of the biotic world, discussed below). In other words, consciousness is an evolutionary transition (Maynard Smith and Szathmáry 1995). In addition, consciousness is an example of biological functionalism, where the function has been selected for as opposed to consciousness itself being the goal (teleological reasoning). Finally, while consciousness can be studied objectively through observation, it has the added quality of subjectivity, which does not exist in the abiotic world. Qualia are the individual instances of a subjective, conscious experience that not only are a feature of living organisms but are *unique* to individual living organisms (Nagel 1974). All these features (emergence, increased complexity, functionality, targets of natural selection, evolutionary transitions, qualia), are embodied in the example of consciousness, and are hallmarks of the abiotic-biotic divide.

The notions of emergence, complexity, and functionalism support the claim that additional principles are required to investigate living systems. Additional support comes from the very definitions of life. An example is the NASA working definition of life referred to in chapter 1, which states that "life is a self-sustaining chemical system capable of Darwinian evolution" (foreword in Joyce 1994). Although this definition is not entirely satisfactory (discussed in chapter 1), it does make the important point that the process of evolution (in this case limited to Darwinian evolution or natural selection) is what distinguishes the living from non-living. The author implies that chemistry alone cannot account for our understanding of what life is. The principle of Darwinian evolution, one of the organizing principles in biology, is also required. Associated with the process of evolution are concepts like fitness, adaptation, natural selection, non-adaptive evolution, complexity, and many others (Sober 2006; Scheiner and Mindell 2020), none of which are adequately appreciated by the first principles approach of physics and chemistry (additional notes 2.5).

Two major laws (or principles) in biology

What are the laws or principles in biology that capture the essential features of living forms, and not found in non-living material? There are at least two, and probably more, very general principles that are necessary to investigate the origin and evolution of life. The first is the principle of natural (including sexual) selection. Lewontin's conditions explain that when there are reproducing units (see additional notes 2.5) that have heritable variation in fitness, the units will evolve by natural selection (Lewontin 1970). This is not to say that natural selection is the only process in evolution, for there are others like genetic drift. But when units within a population are reproducing and they have heritable traits that vary in their effects on fitness, there will be evolution by natural selection. The reason this first principle is such a distinguishing feature of life is because it so clearly does not exist in non-living systems. Not only does natural selection have no meaning in the abiotic world but the phenomenon of natural selection itself emerges out of the collective. Natural selection is itself an emergent phenomenon because measures of fitness are only ever meaningful when one considers them in relation to others in the population. The concept of fitness is a relative one. Using the number of offspring (this is often the most direct way) as a measure of fitness: if one unit produces two offspring and a second unit produces four, one would say the second is fitter than the first, all else being equal. If there is only one unit and nothing with which to compare it, there is no way of knowing just how fit the unit is, at least not until there are others in the population.

The second principle in biology that differentiates biota from abiota concerns the concepts of emergence, complexity, and functionalism introduced above. The term *complexity* is, as with so many other terms in evolutionary biology, loaded with different meanings, contexts, and interpretations (Adami 2002). It is used intuitively here. There is a fundamental difference in the complexity of living and non-living material. In the latter, elements, molecules, and compounds are more, or less, complex because of their chemical structures and the interactions between them. This is also applicable to living systems (as mentioned above, life is ultimately made up of the same elements and compounds as non-life), but in the biotic world complexity takes on additional meanings. One of these is that the living world is comprised of hierarchical levels of organization (Maynard Smith and Szathmáry 1995). Due to their functionality, new kinds of more complex biological units of life emerge from less complex ones. Groups of genetic elements formed genomes, which are compartmentalized in prokaryote cells. In the Proterozoic eon eukaryotes emerged from the cooperation between different types of prokaryote ancestors. Unicellular eukaryotes gave rise to groups of cells and eventually multicellular life, some of which, like the hymenopteran and isopteran insects, are organized into obligatory eusocial communities. The hierarchical levels are formed by units at the lower level, for example prokaryote cells, cooperating to form a new kind of functionally integrated unit, like the eukaryote cell. This kind of transition in complexity does not occur in abiota. The history of this increasing complexity on a macro-evolutionary scale is unique to living systems and appreciating it fully requires more than a reductionist methodology.

3
The theory of life's origins

Theoretical studies (here I include anything that is not "wet laboratory"–based work, such as philosophy, mathematical models, computational simulations, genomics, and bioinformatics) of life's origins have a rich history. Despite the advances, there has been a noticeable disconnect between the goals and findings of theorists and empiricists (Radzvilavicius and Blackstone 2018). Theorists have concentrated on philosophical, mathematical, and fundamental evolutionary questions while the empiricists have focused on the mechanistic issues. Consequently, there have been ground-breaking advances in the mathematical and computational areas that have largely been ignored by experimentalists and not tested empirically. The corollary is also true: wet-laboratory experiments are sometimes developed that are out of keeping with the theoretical imperatives. In many cases, it may not be possible to test many of the theoretical predictions empirically in the laboratory, possibly because they are intractable to experimental manipulation or a reductionist approach. Despite these limitations, theoretical studies have provided answers to some of the key questions and there have been a great number of very significant contributions. In this chapter some of the broad areas in which theoretical insights have made important advances are discussed. These include (i) the theory concerning replicators and replication errors, (ii) issues of group selection at the origin of life, and (iii) the importance of compartmentalization and the concept of the simplest theoretical living system. Metabolism, as discussed earlier, was key for the maintenance of the first living system by generating important biomolecules, but the focus here is replication since it is at this point when birth and death rates existed and natural selection and life history evolution began (Michod 1983). From an evolutionary standpoint, replication is the central issue and metabolism will be discussed further only insofar as it affects reproduction. In addition to the three issues above, a discussion of some of the philosophical questions as well as the advances made from genomic and bioinformatics analyses is warranted to compile a synthesis for life's origin.

The error catastrophe problem

At the very beginning, long before cells existed, there were only replicating molecules like the RNA polymers discussed in the next few chapters. The accuracy or fidelity of the replicative process presented a special problem. The replication of genetic material is never perfect, and copies would not have been identical. This, itself, is not necessarily problematic—errors in replication give rise to the variation that occurs in populations and is the material for evolution by natural selection—but at the origin of life, the fidelity of replication was prohibitively poor. The reason for this is that the very first replicative mechanisms would not have involved the current-day protein enzymes that exhibit high-fidelity copying of nucleic acid (DNA and RNA) sequences. Proteins (at least as we understand them today) did not yet exist. The inefficient copying mechanism prior to enzymatic catalysis became known among theorists as the error threshold problem and constitutes one of the most puzzling questions in the study of the origin of life. It is thought that the error threshold limited the size of the first self-replicating molecules (replication would have occurred passively via base-pairing between strands; fig. 2) to short polymers much fewer than 100 nucleotides. Even the earliest living systems, however, required much longer molecules to develop the necessary genetic potential for life to become sustainable. This problem is handled in living cells by enzymes that replicate nucleotide polymers with great accuracy, and repair mutations when they occur. The large polynucleotide molecules must, of course, encode the very enzymes that replicate them and therein lies the problem, which was explored theoretically by Eigen and others (Eigen 1971; Eigen and Schuster 1977). The upshot of this became known as Eigen's paradox: without protein enzymes the maximum size of a passively replicating molecule that maintains sequence fidelity is less than 100 base pairs, but a polymer of much more than 100 nucleotides is required to code for enzymes.

This chicken-and-egg question (one of many in biology) has, to some degree at least, still not been answered entirely, although there are ways proposed of getting around the problem (see below and chapter 5). Understanding how the problem of genetic deterioration and extinction due to the poor quality of replication was overcome, has been a contentious issue. Some of the first work to address this impasse was that of Eigen

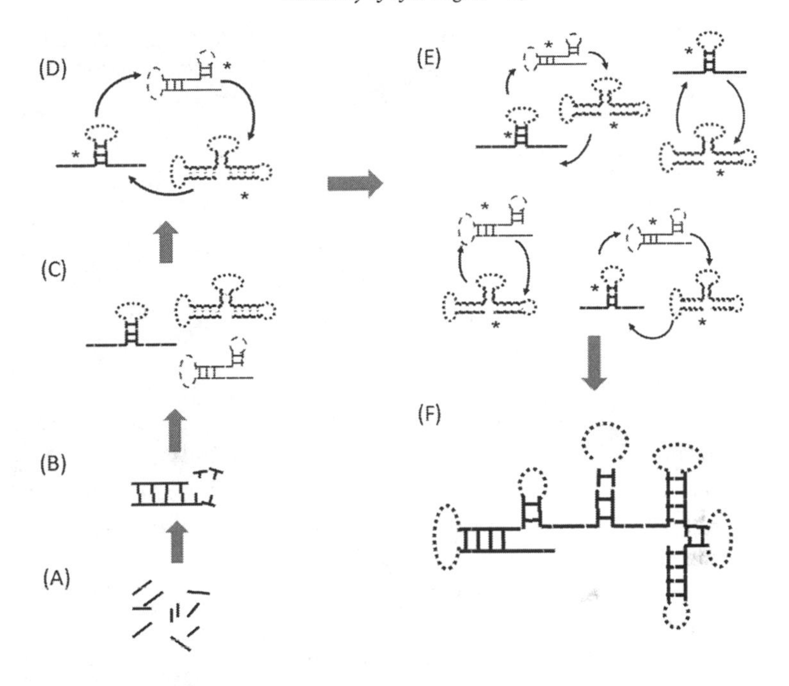

FIG. 2. Hypercycles and the evolution of complex molecular systems at the origin of life. The schematic (modified from Durand and Michod 2010) indicates how complex molecular systems may have evolved. (a) Pre-biotic molecules give rise to polymers. (b) At first polymers would have replicated passively, but due to their secondary structures and their interactions, catalytic functions like replication could emerge. (c) In a population of catalytic molecules there was variation in fitness (the catalytic molecules are shown here as single molecules, but in reality, they were groups of interacting molecules, called LRUs or lower-level replication units; see chapter 6). (d) Replicators can reproduce themselves (indicated by *) but can also promote the replication of others (arrow), which is the basis for a hypercycle. (e) Different hypercycles exhibit different properties and if the conditions are right, compete against each other. (f) More integrated, complex biomolecules emerge from functionally integrated hypercycles.

and Schuster (Eigen and Schuster 1977) who found that, mathematically at least, it was possible for the issue to be addressed by populations of quasispecies (additional notes 3.1). In this scenario, the error rate is so high and the mutation back and forth between parent-offspring states so likely that the fitness of individuals in the population becomes almost irrelevant. Even if a particular mutation is detrimental, it does not have any lasting effect because that genotype's offspring are never accurate copies with the same fitness. In effect, the fitness of the population becomes the central issue. The replicative potential of the population of molecules as a collective becomes much more important than that of the individuals within the population. Within a particular group of quasispecies there is a "master" replicator with a cloud of related individuals (the intensely debated question of group selection at the origin of life is covered in chapter 6 and given a more detailed treatment in chapter 13 in part 2). In Eigen's mathematical argument the connectedness of the population of quasispecies becomes the important factor and the fidelity of the genotype is encapsulated by the population rather than individuals. The distribution of quasispecies in the population helped solve the question of survivability in the face of high mutation rates. In other words, the quasispecies concept overcame the error catastrophe problem by allowing natural selection to act at a population (or group) level. One of the limitations that emerged from this concept, however, was the amount of genetic information that could be contained within a population of quasispecies with polymer lengths <100 nucleotides. There was insufficient genetic potential for the more complex molecules that were important for life to evolve. Eigen and Schuster addressed this second issue with the theory of hypercycles (Eigen and Schuster 1977).

The hypercycle model consists of a cycle of altruistic molecules where, in addition to copying themselves, the replicators promote the replication of others (fig. 2). All the molecules are linked such that each of them catalyzes the creation of its successor, with the last molecule catalyzing the first one. In such a manner, the cycle reinforces itself and the result is a new level of self-organization that incorporates both altruistic (replicating others) and selfish (replicating self) behavior. The coexistence of many genetically unrelated molecules makes it possible to maintain a high genetic diversity in the population, which is one of the proposed solutions to the problem of limited genetic potential. Moreover, it has been shown that hypercycles originate naturally and that incorporating new molecules can extend them, including more and more genetic information. Hypercycles, if the conditions are right, are also subject to evolution and can undergo a selection process. As a result, not only does the system gain information (by increasing in size), but its information content and fitness can be improved (by natural selection). Over time, more complex hypercycles with more members emerge without the need for them to encode large enzymes themselves. The theories of quasispecies and hypercycles were instrumental in developing a system for addressing the problem of replication error and genetic diversity. There are other, and possibly more likely, scenarios for solving this problem (see the next section), but historically the works of Eigen and Schuster stand out as a major conceptual and theoretical advance.

The selection of molecular collectives

The second broad area in which significant theoretical advances contributed to our understanding of the origin of life is the field of group selection and group dynamics. As alluded to above, group selection has been (and remains) an intensely debated topic (Borrello 2005; Eldakar and Wilson 2011; Sober 2009). Sometimes the issue is simply a question of semantics because different researchers sometimes mean different things (West, Griffin, and Gardner 2007) (chapter 13). However, its explanatory potential for some of the major events in the history of life on Earth cannot be ignored. This is the case for life's origins in the RNA world hypothesis (the RNA world, discussed later, is a hypothetical one that holds great appeal for researchers because of its parsimonious nature). At some stage before integrated molecular networks, chromosomes, or genomes existed, there were the replicators above that formed clouds of quasispecies and components of hypercycles. There was always the risk of this system collapsing, however, because of parasitic invaders. The presence of exclusively selfish molecules that invest only in their own reproduction and do not assist the replication of others would have caused the entire hypercycle to collapse. To overcome this, the hypercycle needs to be resistant to invasion by parasitic molecules. One way to achieve this is by compartmentalization (the third major theoretical advance discussed in the next section) or by allowing for a situation where groups or populations of molecules (hypercycles or otherwise) compete against each other. A group with parasitic elements is less fit than one without and, under the right conditions, over time a group of molecules could become resistant to molecular parasites, functionally integrated and eventually indivisible as a unit of selection. Such a system has been captured theoretically by group selection models, such as those developed by Michod (Michod 1983, 1999) and Szathmáry and Demeter (Szathmáry and Demeter 1987). Using the Stochastic Corrector Model (SCM) as an example (Szathmáry and Demeter 1987), the molecules (or templates, as they are referred to in the model) are not replicatively connected. They are, however, grouped into protocells and their functions contribute to the fitness of the protocell. Although the templates can compete for resources, the necessary conditions exist so that selection between groups overrides competition between templates within groups. Protocells with optimal proportions of different templates are fittest and dominate the population. The SCM protects against the problem of invading parasites because replication between templates is not connected and groups are structured in protocells. The group selection basis for the SCM highlights one of the important theoretical advances that permeates through so many of the concepts and models in the origin of life field. Irrespective of how the origin of life problem is investigated theoretically, at some stage when life first emerged there was almost certainly selection at the level of groups (the term *group* is used generically and may be protocells comprising replicator templates, structured populations of hypercycles, or clouds of quasispecies).

Compartmentalization and the chemoton

Selection at the level of populations, groups, or clouds of quasispecies was one of the key events required for life to take hold. However, one of the issues that arises when group selection is invoked is how the groups are defined or structured. It was realized very early that population structure was an important aspect (Michod 1983), and in many of the theoretical models, group structures are either assumed implicitly or explicitly defined. In the quasispecies concept, groups are clouds of sequences related to the "master" sequence. In hypercycles, the groups are replicatively connected molecules, and in the SCM, groups are structurally compartmentalized. Water droplets (Towe 1981; Woese 1980) or microscopic rock crevices (van Holde 1980) could have formed the basis for population structures and the adsorption to mineral surfaces may have been important for the earliest ribozymes to find templates in the immediate neighborhood (Ferris and Ertem 1993). The SCM makes use of the lipid world scenario (Segre et al. 2001) to create boundaries between vesicles (protocells) as a way of structuring groups (Szathmáry 2006). Iron casings (Miller and Bada 1988; Russell and Hall 1997) or gas bubbles (Morasch et al. 2019) could have served similar functions. Compartmentalization is the generic terminology that loosely refers to the segregation of molecules into population or group structures. As indicated, there were many reasons for this important development, including clustering of related genotypes, protection against molecular parasites, increasing the concentrations of substrates and products for enzymatic reactions, and enforcing cooperation between replicators. Recent work has suggested that physical compartmentalization may not have been as important as initially thought. Ribozymes may associate with each other in a coevolutionary process without the need for physical structures to define the populations or groups on which natural selection could act (Levin, Gandon, and West 2020). The upshot, however, is that molecular collectives were still selected for, whether they were structurally isolated or functionally integrated.

One of the earliest conceptualizations of what the simplest compartmentalized living system might have comprised was the idea of a "chemo-

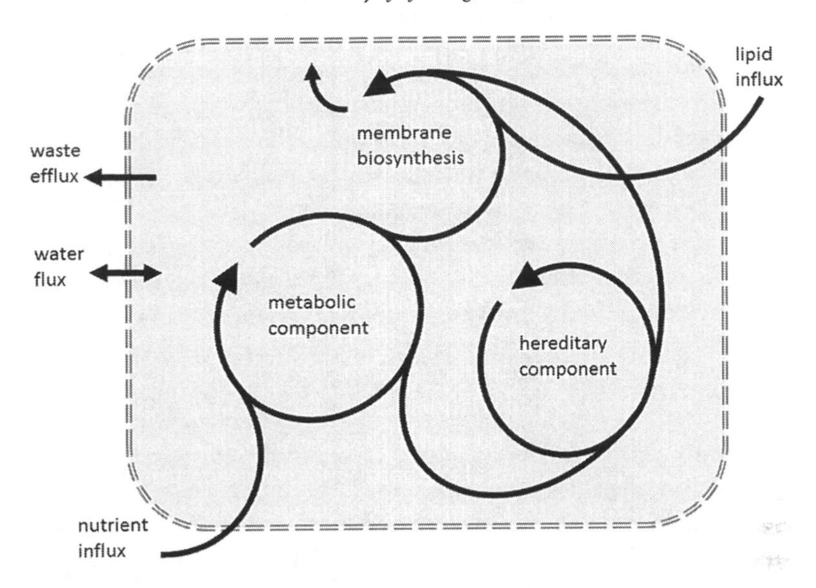

FIG. 3. The chemoton. The chemoton, an abstract protocell developed by Gánti, includes the three essential components of a living system. These are metabolic and hereditary components and membrane biosynthesis, which is important for compartmentalization. The components are stoichiometrically connected and form a chemically sustainable model cell with emergent properties capable of reproduction and evolution by natural selection (Gánti 1975, 2003a, b). Figure redrawn and modified from various sources (for example, Bich et al. 2015).

ton" developed by Gánti (Gánti 1975, 2003a, b). The chemoton (fig. 3) is an abstract model that captures the idea of the simplest living protocell. It has an enduring appeal for origin of life researchers (Griesemer 2015) because it brings together many of the philosophical, mathematical, and evolutionary questions regarding metabolism, hereditary, and replication issues, as well as compartmentalization into a fluid chemical system. The chemoton delivered a chemical perspective on the organization and cyclic stoichiometric relations of a simple biological system and while it is divorced from many of the theoretical data discussed above, it highlights the importance of compartmentalization of the genetic and metabolic features required for early life to begin. Cycle stoichiometry is a feature of biochemical systems and Gánti used his extensive knowledge in this field to describe the kinetics and dynamics of a chemically active, self-reproducing system. The chemoton satisfied what Gánti called the "absolute life" criteria (unity, information capacity, program control, metabolism, and an inherent stability) and the "potential life" criteria (reproduction, hereditary change, and evolution). It comprises three basic autocatalytic chemical components: an information-carrying hereditary system and a metabolic system enclosed in a membrane. These three components were combined stoichiometrically to form a self-replicating autocatalytic cycle, which provided a welcome conceptual bridge between chemistry and evolution. The chemical cycle has become a heuristic for evaluating many, if not all, of the biochemical origin-of-life questions. At the same time, the reproduction and hereditary nature of the information-bearing chemoton are not contrived. These features, so necessary for its evolution, are emergent, self-organizing properties that were not built into the model and are fundamental to cellular life (Wedlich-Soldner and Betz 2018; Wolkenhauer and Hofmeyr 2007).

The hereditary information, and the preservation of its integrity, is one of the core elements common to all the theoretical insights, and the hereditary nature is one of the essential criteria in any of the conceptualizations of the first living system that evolved by natural selection (see Lewontin's conditions in chapter 2). Developments in the fields of computational biology and bioinformatics provide unprecedented access to the hereditary material of life, and the contributions from these fields should be integrated into the models and discussions of life's beginnings.

Evolutionary genomics and computational simulations

One of the distinguishing features of the contemporary era in biology is the generation and analyses of massive amounts of genomic data. The vast amounts of data and sophisticated computational and mathematical methodologies for analyzing them have allowed biologists to examine features of the earliest putative genomes that were previously inaccessible. It is generally accepted that, since all cells use the same D-sugars and L-amino acids, share homologous ribosomal genes, have the same basic genetic code, and synthesize proteins via the same mechanisms, all life forms share a common ancestor(s). The comparative analyses of genome data assist with tracing this ancestry all the way to the earliest living organisms.

In a paradigm-shifting analysis, Woese and others compared the sequence data of ribosomal genes, some of the most slowly evolving and ancient of all genes, and found that all life on Earth emerged from two fundamentally different kinds of prokaryotes, the archaebacteria and the eubacteria (sometimes referred to as the true bacteria) (Woese and Fox 1977). Prior to Woese's works the evolutionary relationships between archaebacteria and eubacteria were unclear. He (additional notes 3.2) discovered that the two divisions of prokaryotes are phylogenetically distinct. The genetic differences, the presence of biochemically unrelated lipids (isoprene ethers or fatty acid esters) in their cell membranes (Wachtershauser 2003), and the enzymatic pathways involved in archaebacterial and eubacterial membrane biogenesis are non-homologous. The proteinaceous components of the cell walls of archaebacteria and eubacteria share even less chemical similarity. The marked differences in these two groups of organisms suggest that the very earliest cells may have been quite different from the kinds of prokaryotes we see today. If the assumption is made that the archaebacteria and eubacteria evolved from a common ancestor, an examination of the similarities between members in the two kingdoms of bacteria gives us a glimpse of what the last universal common ancestor (LUCA) may have looked like. Of course, an awful lot happened between the first replicators and groups of integrated molecules and the emergence of LUCA (chapters 6 and 7), but the discovery that the two prokaryote kingdoms evolved from some other kind of ancestral organism that was at the root of the tree of life was, at the time, a stunning revelation. What was this ancestral organism and is it possible to identify any of its characteristics? A closer examination of what the putative LUCA may have been, yielded some surprising results.

The field of genomics provides the molecular sequence data used to recapitulate the evolutionary histories of the archaea and eubacteria. It also presents an opportunity to ask even more fundamental questions about the genomic origins of the very first cellular organisms. Koonin and others have investigated whether comparative genomic data are able to resolve the origins of the very first organism with a rudimentary genome. They asked whether it is possible to characterize the primitive genome of LUCA (Koonin 2003, 2009; Koonin and Martin 2005; Koonin, Wolf, and Puigbo 2009). Some profound insights emerged. Looking at the phylogenetic consistency of NUTs (nearly universal trees are phylogenetic trees that contain all the orthologous gene clusters of a representative sample of the prokaryote genomic data), it was found that although the topologies of NUTs exhibit variation, they cluster in the region of a consensus tree. NUTs do indeed converge on a centroid, which represents the hypothetical root of the tree of life (ToL) and support the concept of a LUCA. That said, the topologies of trees for prokaryotic genes are so diverse that it is more appropriate to describe the evolution of bacteria and archaea as an FoL (forest of life) (Koonin, Wolf, and Puigbo 2009) in which a central trend can be discerned, rather than a single ToL. The reason for this is because HGT (horizontal gene transfer) is so pervasive in prokaryotes and was probably even more so at the early stages of the evolution of life when compartmentalization may not have been complete, that a complete genomic characterization of the LUCA is not possible. Thus, we can confidently place into LUCA only a relatively small set of genes that are the least prone to HGT. These are, primarily, those for components of the translation and, to some extent, transcription systems, and we can satisfy ourselves only with tentative, probabilistic inferences regarding the rest of the LUCA's gene repertoire.

It is important to emphasize once again that LUCA is far removed from the origin of life and itself was undoubtedly the product of extensive earlier evolution that included massive HGT. Koonin went on to show that while there is a phylogenetic signal of life's origin, it is not a singularity. Prior to the existence of prokaryotes as we know them today (Koonin 2016), there was likely extensive gene exchange between compartments of genes. It is not, in fact, correct to talk about a single origin of life. Using this methodological approach, genomicists have arrived at a conclusion that is consistent with the philosophy and theory covered in the first three chapters. For philosophers of biology, the origin of life is captured more by the evolutionary process than a substance or a specific event. Theorists demonstrated that scientific pluralism, rather than a single categorical explanation, is more helpful to describe the emergence of the first living systems. These philosophical and theoretical advances are echoed by the genomic findings, which suggest a collection of information-exchanging protocells rather than a single LUCA at the origin of life. The next question we can ask is this: how much of the philosophy, theory, and genomics has been investigated experimentally? This is the basic question that will occupy most of the succeeding chapters before a synthesis for life's beginning is proposed.

4

Life at the very beginning I

THE CHEMISTRY OF THE FIRST BIOMOLECULES

From an evolutionary biology perspective, and detailed in the previous chapters, the criteria for the simplest molecular system to be considered alive and a target for natural selection, are that the collection of molecules is a unit with the capacity for reproduction and heritable variation in fitness. Theoretical studies typically refer to these early molecules (whether one molecule or a collection of molecules that replicate as a unit) as replicators. As the reader is by now aware, terminology and the meanings of evolutionary terms are an important consideration. The term *replicator* is an abstract one used generically. It sometimes refers to a single molecule and at other times to a collection of molecules that replicate as a unit. For empirical studies and to develop a mechanistic understanding we must ask: What exactly were these replicators? And what experimental system can be used in the laboratory to study their properties and behavior?

Considering that all living material comprises DNA, RNA, and protein, it is reasonable to assume that one or more of these molecules (or at least variants of them) played a central role at the very beginning. They can be incorporated into the exemplars of the theoretical replicators. For several reasons, it is claimed that RNA, or something like it, was the first molecule that included hereditary genotypic information and due to its enzymatic activity could catalyze its own or others' replication. This assumption is inherent in the "RNA world hypothesis" and there is a substantial body of work to support this theory (for recent reviews see Higgs and Lehman 2015; Robertson and Joyce 2012). However, at the outset there are several caveats and points to note. It is possible, if not probable, that the first replicators were not RNA or any other extant biomolecule (fig. 4). RNA and its nucleobases are vulnerable to chemical degradation and it has been proposed that a pre-RNA world existed that no longer exists today (Bohler, Nielsen, and Orgel 1995). An RNA world could have emerged from this

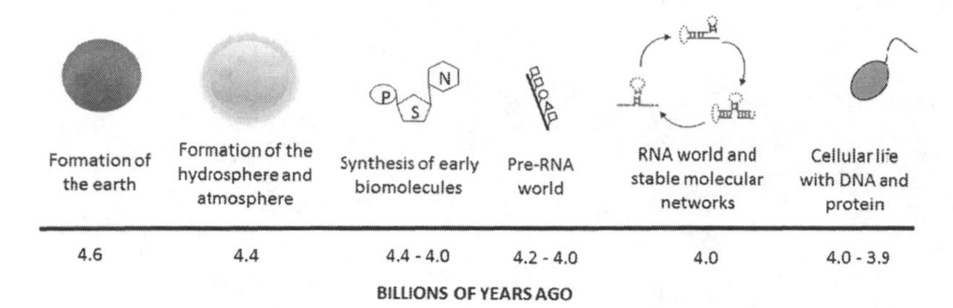

FIG. 4. Major steps in the origin of life. The timescale is a gross approximation. The synthesis of early biomolecules includes nucleobases (N), sugars (S), and phosphate moieties (P). The schematic is redrawn and modified from Joyce 2002.

pre-RNA world once the chemical conditions were more suitable for its viability (for example, a buffered environment or compartmentalization that protected RNA molecules from the environment).

The concept of the RNA world is a parsimonious one. It provides an important conceptual and biochemical bridge from single molecules with both genotypic and phenotypic information to the current world where genotype and phenotype are separated (DNA and protein). For example, ribozymes are RNA molecules with hereditary potential and at the same time have catalytic activity. Genotype and phenotype are one and the same. In principle at least, a living system based only on RNA has no need for proteins or DNA. In addition, the catalytic repertoire of ribozymes includes functions like polymerization, ligation, and recombination, which would have been essential for the replication and organization of RNA strands. This same duality does not exist, or certainly not nearly to the same degree, in DNA or protein molecules. Nevertheless, while the RNA world is appealing because of the parsimonious nature, it is unlikely that it emerged and was completely functional as a stable living system prior to the emergence of DNA, the genetic code, amino acids, or small peptides. Amino acids almost certainly existed at the same time as RNA and it is reasonable to assume that from the pre-RNA world, an RNA one emerged that coevolved with small peptides (Wills and Carter 2018) and possibly even DNA. There are alternative views and a significant number of researchers are not in favor of the RNA-world scenario. They present viable and reasonable alternatives (for example Guseva, Zuckermann, and Dill 2017). But as things stand and given our current understanding, on balance I favor the RNA-world hypothesis as a model system. Perhaps it is most appropriate, because of all the limitations and gaps in our understanding, that we consider the RNA-world hypothesis "the worst explanation we have for the origin of life—except for all the others" (Bernhardt 2012). Given that many origin of life researchers are suspicious of the RNA-world scenario, it is worthwhile detailing some of the chemistry that supports the emergence of the molecular substructures and RNA polymers in the prebiotic world and highlighting the strengths and weaknesses of the theory. Only some of the basic chemistry is covered and other issues like the energy sources that drove the synthesis of large molecules, or the early Earth's environmental conditions, are not discussed.

Could the molecular building blocks of RNA have emerged on the early Earth?

Is it reasonable to assume that the biomolecules (the nucleobases, ribose sugars, and nucleotides) that make up RNA emerged from the mix of gases, minerals, chemically active compounds, and small organic molecules that existed on the primitive Earth? This does seem reasonable, yes, even if we are a long way off from demonstrating exactly how this happened. In a series of what were at the time ground-breaking chemical experiments, it was demonstrated that some of the simple biomolecules that formed the building blocks of RNA could have emerged, but only under a set of very specific conditions. It is also possible that some of the important components arrived from other parts of the solar system. Amino acids and nucleotides have been discovered on meteorites (for example, the Murchison meteorite), indicating that some of the important biomolecules may not have emerged spontaneously on Earth. This extra-terrestrial origin of some biomolecules is important because the spontaneous emergence of some compounds in the primitive atmosphere depended upon the presence of specific gases in specific quantities, and even then, not all the requisite amino acids may have been produced. Furthermore, the reducing atmosphere necessary for the formation of some of the early biomolecules was questioned (Draganić 2005). It seems the atmosphere may have been too oxygen-rich for the proposed amino acids to form (Frost et al. 2008), and many planetary scientists now argue that the reducing gases so essential to the above experiments should not be invoked (Canil 2002). Nevertheless, the research that followed the Miller-Urey experiments, which demonstrated the formation of some of the current amino acids (Miller 1953; Miller and Urey 1959a, b), even if their experimental conditions were unrealistic, were a prelude to many other ground-breaking works and set the tone for investigating the emergence of the RNA components. Oró and others have generated nucleobases from the reactants that would have been present in the early atmosphere (for example, Oró 1960, and see additional notes 4.1) and Butlerov discovered that mineral catalysts can polymerize the formation of various sugars from formaldehyde (the formose reaction) (Leicester 1940). Eschenmoser, Orgel, and others expanded on this work to demonstrate how greater amounts of ribose (the sugar component in RNA) may have arisen so that an RNA world could take hold (Drenkard, Ferris, and Eschenmoser 1990; Orgel 2004). Nucleosides are comprised of ribose and nucleobases, and the formation of adenosine (adenine + ribose) was demonstrated by Fuller and others (Fuller, Sanchez, and Orgel 1972). Nucleosides comprising the other nucleobases (uracil, cytosine, and guanine) have been more problematic. However, routes to their formation via alternate nucleobases are possible (this is one of the reasons advanced for the existence of a pre-RNA world).

The next question was to determine whether nucleotides (nucleosides + phosphate) could have emerged from nucleosides. For a long time concentrations of phosphate were considered a significant limitation; however, a very recent study demonstrated a "carbonate-rich lake solution to the phosphate problem" (Toner and Catling 2020). Phosphate-rich lakes could have formed on the prebiotic Earth because of carbonic acid weathering in the rich CO_2 atmosphere. The emergence of nucleotides has been just as difficult to pin down as the formation of nucleosides (Pasek et al. 2013), although Lorhmann and Orgel demonstrated efficient phosphorylation by drying and heating nucleosides with acidic ammonium phosphate minerals in the presence of urea, which functioned as a catalyst (Lohrmann and Orgel 1971). This reaction produced a complex mixture of phosphorylated products. The attempts to direct this reaction to the synthesis of, particularly, nucleoside-5′-phosphate or 5′-triphosphate have had some success and nucleoside phosphorylation in good yields was also shown using calcium phosphate minerals (hydroxylapatite). In addition to using phosphates, prebiotic chemists also investigated more reactive forms of phosphorus, such as inorganic phosphate, anhydrides of phosphate (like pyrophosphate or metaphosphate), and reduced forms of phosphorus (like phosphite). These phosphorous species were shown to result from the interaction of iron-rich meteorites with water (Pasek et al. 2013). A reaction of nucleosides with tri-metaphosphates in strongly alkaline conditions formed nucleoside 2′-3′-cyclic phosphates that hydrolyze to a mixture of nucleoside 2′- and 3′-phosphates (for example, Saffhill 1970). An alternative route to nucleotide synthesis that bypasses problematic piece-wise assembly from ribose sugar and nucleobases was demonstrated by Powner and others (Powner, Gerland, and Sutherland 2009). In this approach, cytosine nucleotides formed from small molecules through a mostly water-based multistep synthesis. Subsequent UV irradiation of cytosine nucleotides formed uracil nucleotides.

Although the de novo generation of ribonucleic acids and their substructures has been achieved by prebiotic chemists, the issue has not been completely resolved. There are difficulties in explaining how sufficient quantities of reactants and products were produced. This is still another reason that leads chemists to believe that RNA may be a product of a pre-RNA world (Orgel 2004). It is suggested that simpler biomolecules that were synthetically less problematic may have played a role prior to the RNA world. Such biomolecules may have included non-ribose nucleic acid analogs like threose nucleic acids (TNAs), peptide nucleic acids (PNAs), and glycol nucleic acids (GNAs). TNAs, for example, form stable Watson-Crick structures with themselves and RNAs, indicating that prior to a pure RNA world it is quite feasible that there was a stage where ribose and non-ribose nucleic acids coexisted. However, the chemical optimality of RNA as an informational and functional molecule and its presence today in all living systems indicates that over time ribonucleotides replaced any other non-ribose nucleotides as the basic information molecule of life.

Despite the gaps and limitations in our understanding of the origin of ribonucleotides the emergence of these molecules is in keeping with the assumptions concerning the geochemistry of the early Earth. The next challenge was to determine how polymers of ribonucleotides arose.

The abiotic synthesis of RNA polymers

The main problem with the polymerization of ribonucleotides in aqueous solution is that it is an uphill reaction (it requires energy) and does not occur spontaneously to any significant extent. It is believed, therefore, that activated derivatives of the ribonucleotides discussed above, such as nucleoside 5′-polyphosphates or nucleoside 5′-phosphorimidazolides, were important for the first polymerization events. This has been borne out experimentally. Three principal nucleophilic groups in activated nucleotides are able to participate in the polymerization reaction (the 5′-phosphate, the 2′-hydroxyl, and the 3′-hydroxyl group). The reaction of a nucleotide or oligonucleotide with an activated nucleotide yielded, in decreasing order of abundance, 5′,5′-pyrophosphate, 2′,5′- phosphodiester, and 3′,5′-phosphodiester-linked adducts (Sulston et al. 1968), and coupling more than 15 activated adenine and uridine-monophosphates can be obtained using lead as a catalyst under eutectic conditions (Kanavarioti, Monnard, and Deamer 2001). In these early experiments the product contained a large proportion of 2′,5′ linkages (instead of the typical 3′,5′ linkages), which at the time was considered problematic for subsequent base-pairing but has since been shown not to be an obstacle.

A more appealing strategy for catalyzing more favorable linkage-specific reactions was via the adsorption of ribonucleotides to mineral surfaces. Surface-enhanced polymerization of nucleoside 5′-phosphorimidazolides and related activated nucleotides have been investigated extensively on the clay mineral, montmorillonite (for example, Ferris and Ertem 1993). It was possible to achieve the polymerization of 40–50 ribonucleotides in which adenines were primarily 3′,5′-linked and pyrimidines 2′,5′-linked (Ferris 2002; Ferris et al. 1996). The two types of bonds result in heterogeneous backbone linkages. As indicated above, however, this did not present any functional constraints (Engelhart, Powner, and Szostak 2013). The formation of 40 nucleotide-long polymers on montmorillonite clay has also been obtained using 1-methyl-adenine and uridine-monophosphates. A detailed analysis of catalysis by montmorillonite suggested that polymerization occurred at a limited number of structurally specific active sites within the interlayers of the clay platelets (Wang and Ferris 2001). These experiments provide a plausible explanation for prebiotic ribonucleotide polymerization and although the conditions were very specific, they reinforce the claim that life could have started on mineral surfaces, perhaps in clay-rich muds. An alternative to this scenario is the dry state polymerization of non-chemically activated ribonucleotides. The acidic form of cyclic 3′,5′-GMP (guanosine monophosphate), for example, autopolymerizes to form polymers of 40 ribonucleotides long if dried at elevated temperatures (Morasch et al. 2014).

Once short polymers of between 40 and 50 ribonucleotides had formed, the question that is of paramount importance before evolution by natural selection can occur is this: how did these molecules reproduce? To the early origin-of-life biochemists, it seemed impossible for molecules this small to have any enzymatic activity that could facilitate replication in any way. Furthermore, molecules longer than 40–50 ribonucleotides could not have been maintained without enzymatic replication (see chapter 3 and the error catastrophe problem). These short polymers must, therefore, have been maintained by passive, nonenzymatic, chemical base-pairing.

Nonenzymatic replication of short RNA polymers

For chemical replication to occur, the synthesis of a complementary ribonucleotide polymer must take place using a preexisting polymer as a template. The basic principle is that a double helical complex is formed when a polymer is incubated with an appropriate mixture of complementary monomers or short oligomers under the necessary chemical conditions. Such a scheme was investigated using ribonucleotide monomers activated with 2-methylimidazolide or 5′-phosphorimidazolide resulting in the relatively efficient formation of a 50 ribonucleotide-long guanine polymer using a cytosine template. The fidelity of complementary base-pairing was investigated in great detail with co-polymers and excesses of particular ribonucleotides (Inoue and Orgel 1983). Incorporation of guanine opposite cytosine in the template was most efficient, while incorporation of uracil opposite adenine was least efficient. Incorporation of adenine opposite uracil or cytosine opposite guanine were of intermediate efficiency. These results suggested that the accuracy of replication was generally good, except for the erroneous incorporation of guanine on some RNA templates because of guanine / uracil wobble pairing. Under more favorable conditions the fidelity of the process can be enhanced. For example, the successful and accurate extension of a primer across adenine-rich sequences was demonstrated in ice eutectics containing metal ions (Monnard and Szostak 2008), and the immobilization of template primer strands with periodic replenishment of activated ribonucleotides greatly improved the copying of all four ribonucleotides even further (Deck, Jauker, and Richert 2011). These data showed that the passive replication and maintenance of a population of 40–50 ribonucleotide-long polymers, which was the starting point for the theoretical work in chapter 3, was chemically feasible.

An alternative scheme for a nonenzymatic self-replication system was devised based on template-directed ligation of activated short 3′,5′-linked oligonucleotides. This was superior to mononucleotides as substrates with respect to regiospecificity and the temperature ranges at which the reactions occurred. Of course, in these instances fidelity is compromised, since single base mismatches are much more likely to occur in an oligomer of several nucleotides than in a mononucleotide (James and Ellington 1999).

The emergence of RNA life

At this stage of the RNA world, the substructures of RNA, ribonucleotides, short polymers of 40–50 ribonucleotides in length, and a popula-

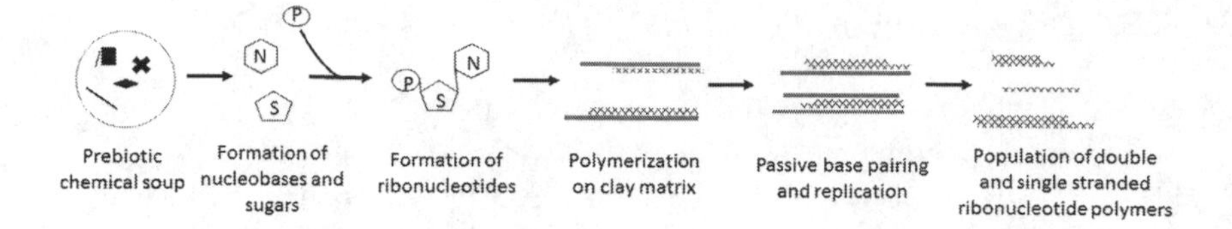

FIG. 5. The formation and replication of short RNA polymers from molecular substructures.

tion of these short polymers existed on the early Earth (fig. 5). For RNA-based life to have taken hold and eventually give rise to all life on Earth, much more complex polymers (>50 ribonucleotides long) were required. How this happened depended on the emergence of catalytic RNA molecules.

5
Life at the very beginning II
THE EMERGENCE OF COMPLEX RNA MOLECULES

There remain gaps in our understanding of how short RNA polymers arose. Nevertheless, it is quite reasonable to assume that molecules of less than about 50 ribonucleotides in length emerged in the prebiotic world. This may have been by spontaneous chemical reactions between reactants that existed at the time and augmented by the physical properties of the environment. The introduction of some chemical components via extraterrestrial meteorites may have contributed. The maintenance of a population of passively replicating molecules of ~50 ribonucleotides in length was also possible. However, before RNA-based life could have evolved, larger, more complex molecules (see additional notes 5.1 for more on the meaning of complexity) with enzymatic potential were required. The improved accuracy of enzyme-mediated copying of RNA molecules was essential. This presents a chicken-and-egg problem, since large molecules (>100 ribonucleotides) cannot be replicated accurately without a molecule that is itself large enough to have enzymatic activity like polymerization (see chapter 3, which discusses the error catastrophe problem and Eigen's paradox). How then did the very "first" complex molecule come into existence? This is the next major question that needs to be answered en route to a complete synthesis for the origin of life.

The essential function required for the persistence of large RNA molecules is enzyme-mediated replication. Furthermore, for the replicating population to be subject to evolution by natural selection there must be heritable variation in fitness (see Lewontin's criteria in chapter 2). For a long time, the holy grail of the RNA world has been the search for a single ribozyme with self-replicating ability, a so-called replicase. There

I am very grateful to Nisha Dhar for her reading, corrections, and comments on this chapter. At the same time, any errors are mine alone.

are at least two problems associated with this endeavor. First, as indicated above (see also chapter 3), how such a molecule could have emerged from much smaller ones is a major obstacle. A replicase would presumably be a large, complex molecule much greater than 50 ribonucleotides. How such a molecule could have emerged from the pool of smaller molecules is problematic. The second issue is whether it is reasonable to think that such a molecule could have existed at all (Durand and Michod 2010). For a molecule to replicate itself the polymer would need to be both unfolded (so that it could serve as a template for replication) and folded into secondary and tertiary structures (so that enzymatic activity was possible). This is theoretically possible if the molecule comprised two domains functioning independently, although such a molecule would be very large indeed. In addition, the two domains would both need to have polymerization potential so that they could copy each other. The structural constraint suggests it is highly improbable that such a molecule could have existed (certainly not at the beginning) unless it was truly unique in terms of its folding thermodynamics and enzymatic potential. To overcome this problem, two molecules that emerged independently such that they could replicate each other reciprocally are required. But this compounds an already difficult problem since now two large, complex molecules are necessary. What is more, they both need to have evolved polymerase activity independently and the polymerization activity would have needed to be highly efficient for accurate replication of the target sequence. The replicase route does not seem feasible. There are, however, a few alternate explanations that bypass the requirement of a replicase for the emergence of complex ribozymes with polymerization and other enzymatic activities.

The emergence of complex ribozymes

There are a few routes by which complex ribozymes could have emerged. As Higgs and Lehman point out, cooperation at the molecular level is a common theme in all these scenarios. Cooperation between molecules was essential for the evolution of replicating sequences (Higgs and Lehman 2015) (see additional notes 5.2 for the meaning of the term *cooperation*). Joyce and others have identified the importance of template-mediated polymerization of oligomers by ribozymes (Joyce 2009). This function involves the chemical joining of RNA oligonucleotides complementary to a template and was achieved by a relatively simple ligase (Bartel and Szostak 1993). Ligation and polymerization reactions are biochemically very similar, so it is easy to see how polymerization activity emerged from ligation activity. Much larger, true polymerase ribozymes were subsequently developed from this simple molecule (Johnston et al. 2001). This was a major experimental breakthrough because it provided a link between a relatively simple ligase and the emergence of a complex polymerase. It was later demonstrated that complex molecules with ligase or polymerase activity could have synthesized new ribozymes, eventually leading to a simple network of two ribozymes in a reciprocal self-sustaining replicating system (Lincoln and Joyce 2009). This scenario comes close to addressing the issue of sustainable replication by complex ribozymes, although there are still some obstacles. The simple ligase from which the system was developed remained too large (> 90 ribonucleotides) to explain its emergence from non-enzymatically replicating polymers. In addition, base-pairing complementarity between the ribozyme and substrate was a necessary feature of the system and the heterogeneous nature of oligonucleotides in the prebiotic world decreases the probability of this occurring.

A second mechanism that may have played a role in the formation of complex molecules was recombination. One such reaction system that led to the formation of an extant self-splicing ribozyme was demonstrated by Hayden and Lehman (Hayden and Lehman 2006). The ribozyme emerged following the assembly of a recombinase that itself formed from the secondary structures of ribonucleotide fragments directed by complementarity between sequences. This system provides a viable alternative to the first scenario above, in the sense that complex molecules emerged from much smaller fragments. The only requirement was that the assembly of fragments that formed a recombinase relied on complementarity. In other words, it was important that fragments bound each other through complementary base-pairing. This may have been possible, although as indicated above the prebiotic environment was abundant in randomized, heterogeneous fragments and this may have constrained the formation of cooperative assemblages.

A third potential path that leads to the formation of complex molecules was demonstrated by Dhar et al. (Dhar et al. 2017). In this proof-of-concept system, larger molecules were obtained via self-ligation reactions in much smaller ribozymes. Truncated forms of a polymerase (with a ligase core) developed by Johnston and others (Johnston et al. 2001) had the ability to join substrates (35 ribonucleotide sequences) to their own ends, thereby increasing in size. This system overcomes some of the mechanistic limitations of the first scenario because the ligase activity was demonstrated without any specific base-pairing between ribozyme and substrate. Of course, that does not mean that the ribozyme-substrate systems were free of complementarities—in a population of nucleotide sequences one cannot predict what molecular potential complexes and secondary structures may form—but there was no intentional base-pairing designed in the experimental system. The products of these ligation events were larger, more complex (at least in terms of sequence length) molecules. The inference is that by joining random sequences to themselves, there is the possibility that one or more of the simple ligase molecules developed polymerase or recombinase activity when it increased in size. An important feature of this simple ligation scenario is that the ligases could tolerate variations in the substrate sequences. Ligation activity did not depend on the type of substrate, which is noteworthy because the prebiotic system would have comprised a heterogeneous pool of oligonucleotides. This is also relevant because, with a diverse pool of substrates being generated passively, the diversity of the products of the ligation activity is enhanced, increasing the likelihood of polymerases or recombinases emerging. The emergence of polymerases or recombinases relies on chance events, but this in itself is not problematic. Many events in the evolutionary history of life depended on probabilities (Koonin 2011; Ramsey and Pence 2016). In addition, the chance of a favorable outcome is increased because of the non-specific nature of the simple ligase and the heterogeneity of the available substrates. The issue of ribozyme specificity revealed further insights into the evolution of complex networks.

Short ligases (45 nucleotides), which could have evolved passively, were functionally promiscuous (Dhar et al. 2017). They were relatively non-specific in the kinds of substrates they ligated. This increases the chance of generating a product with biologically relevant activity like polymerization or recombination. It was also found that the larger, more complex ligases (which themselves could have been the product of ligation events) were more specific in their activity. In other words, they were more selective in the kinds of substrates they could ligate. This inverse relationship between molecular complexity and functional flexibility (fig. 6), which is inherent in the molecule's biochemistry (see additional notes 5.3), was important for the emerging complexity. Early in the process when only simple ligases existed, functionality was flexible, allowing for many of the heterogeneous substrates to be joined to the small ligases. Out of this molecular pool could emerge larger, more complex molecules with more specific activity. The larger molecules would have been unable to join substrates, which may have adversely affected their function. In addition, these larger molecules could have exhibited specialized functions like polymerization or recombination from which self-sustaining replication systems evolved.

The three scenarios described above, or more likely a combination of

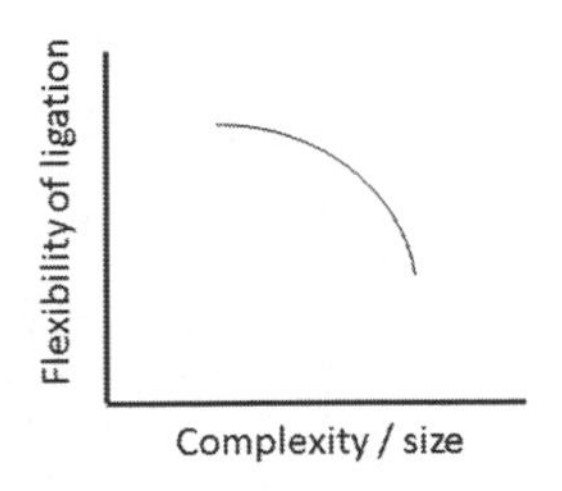

FIG. 6. A molecular trade-off in the earliest ligase ribozyme. A trade-off exists when there is a causal relationship between two traits that exhibit an inverse correlation. Here, the relationship is between molecular size (sequence length of the ribozyme) and flexibility (the kinds of substrates the molecule ligates). Small ligases are more flexible in the kinds of substrates they can use. Larger ligases are less flexible (Dhar 2016).

them, can explain how it was possible for large, complex ribozymes to emerge from small ligases. A few points are worth making. The possible events discussed in this chapter describe how the initial steps of increasing molecular complexity may have occurred. There is no way of knowing exactly what happened. What is more, without knowing which biomolecules existed at the very beginning, it is impractical to imagine and work through all the possible steps. But with our current understanding of ribozymes, we have a sense of how it *may* have happened. In addition, we can appreciate that many of these initial events relied on chance, which as Koonin (Koonin 2011), Ramsey and Pence (Ramsey and Pence 2016), and many others have discussed, plays a significant role in all evolutionary events. Nevertheless, once a self-sustaining replication system with heritable variation in fitness emerged, evolution by natural selection would have impacted subsequent events at the origin of life, which is the issue that needs to be considered next.

6

The origin of life was an evolutionary transition in individuality

The history of the living world can be described as a series of evolutionary transitions (ETs) (Maynard Smith and Szathmáry 1995). The very first of these was the origin of life itself and the theoretical insights concerning ETs have been applied to the emergence of the first molecules capable of self-replication (for example, Szathmáry and Demeter 1987; Michod 1983). Evolutionary transitions in individuality or ETIs (a subset of ETs, discussed below) are conceived as the transition from groups of individuals to new kinds of individuals (for example, from cell groups to a new multicellular organism). But at the origin of life, there were no evolutionary individuals to begin with. In other ETIs, such as the evolution of multicellularity, the cellular slime molds (Olive and Stoianovitch 1975), the volvocine algae (Kirk 2005), and other lineages are examples of experimental model systems for investigating the steps from unicellular organisms to multicellular ones. But at the very beginning of life, it is not clear what precisely the first "living" molecules were. What were the beginning entities that can be used to study the origin of life as an evolutionary transition in individuality? To work around some of these limitations, evolutionists have relied largely upon theoretical models (see chapter 3), which provided many of the key advances. For example, theoretical work has illustrated that, regardless of what the earliest replicating molecules might have been, cooperation between them and selection of collectives in some form or another must have played a key role for life to take hold (Higgs and Lehman 2015; Michod 1983; Levin, Gandon, and West 2020) (the issue of selection at the level of collectives is dealt with in chapters 13 and 15).

Investigating the transition from non-living to living material presents

I am very grateful to Rick Michod for his reading, corrections, and comments on this chapter. At the same time, any errors are mine alone.

unique challenges. It has been difficult to translate some of the concepts and results from the theoretical works concerning the origin of life into laboratory model systems that are tractable to experimentation. The concept of the molecular "replicator," for example, is usually dealt with abstractly in mathematical models and often discussed without taking into account the molecular constraints inherent in the biochemical mechanisms (Durand and Michod 2010). A knowledge of the biochemical parameters and constraints, however, complements, refines, and limits what the theoretical findings can predict. Variables like the rate of replication, availability of substrates, and population structures of replicators all depend on molecular mechanisms that constrain the modeling outcomes. Which engineered molecules and model systems can be used to obtain these kinds of data?

In some instances, the gene is the perceived replicator, in some a single ribozyme-like molecule (the replicase), and others refer loosely to "primordial replicators" (Krakauer and Sasaki 2002) without saying what they were. But using molecules like genes or single ribozymes as exemplars of the origin of life replicators is problematic. Genes as we understand them today did not exist at the very beginning and the biochemical data, discussed in the earlier chapters, suggest that a hypothetical ribozyme replicator such as a "replicase" did not exist. Nevertheless, it is necessary that we harmonize the theoretical concepts of a replicator with the current state of knowledge of the biochemistry of the earliest molecules. To do so, the origin of life must be conceptualized by focusing more on the process as opposed to a discrete event, or the instantiation of a single "living" molecule (see chapters 2, 3, and 5).

An interpretive view of the origin of life as an evolutionary transition in individuality

The general conception of evolutionary transitions (ETs) (Maynard Smith and Szathmáry 1995) has been revised and reworked by several researchers (for example, Hanschen et al. 2018; Koonin 2007; Szathmáry 2015) with the evolutionary transitions in individuality (ETIs) (Michod 1997, 2007; Michod and Roze 1997; West et al. 2015) being a subset of these. The terminology can be confusing, even to specialists in the field, but superficially the ETs include transitions where there is a major shift in complexity (what exactly complexity is, is discussed in chapters 2 and 3). What constitutes a major or minor shift in complexity can be debated, but ETs include events like the origin of life, the evolution of photosynthesis, or the emergence of language and consciousness. ETs may also include ETIs, but ETIs have a much more specific meaning. ETIs deal with transitions where individuals that were capable of independent replication before the transition become functionally integrated such that they can replicate only as part of a new kind of individual after the transition. In ETIs groups of individuals evolve into new kinds of individuals (Michod 1999, 2005). Questions concerning the units and targets of selection (Okasha 2006), and the fitness associated with units at different levels of selection (Michod 1999), have special relevance for ETIs (for more on units, levels, and what is meant by an "individual" see additional notes 6.1).

The general theory of ETIs concerns how groups of one kind of individual become a new kind of individual (Michod 1999, 2005; Hanschen et al. 2015, 2018). There is a transition from a lower level of organization to a higher one. Broadly speaking, the process occurs via several stages (fig. 7). At the lower level there is a population of individuals that interact and evolve in the same way as any other population of individuals where there is heritable variation in fitness. If the conditions and circumstances allow, individuals may form aggregations or groups. Iterative cycles of cooperation, conflict, and conflict mediation (Michod and Nedelcu 2003) between members leads to the emergence of a group dynamic such that the group itself can be selected for by natural selection and evolve novel adaptations at the group level of organization (Michod 1999). Group-level selection (see chapters 13 and 15 for what is meant by group selection in this book and West, Griffin, and Gardner 2007 for a thorough treatment of the terminology) facilitates the evolution of functionally integrated groups that emerged from individuals at the lower level. Over time, the collectives become new kinds of individuals that are capable of reproduction only as an indivisible unit. Examples of ETIs and the ones that have been investigated more intensely are the origins of the eukaryote cell, multicellularity, and eusocial insects. The transition from unicellularity to multicellularity is, perhaps, the easiest of the ETIs for a non-specialist to imagine.

The origin of life can be studied as either an ET or an ETI. The origin of life as an ET refers to the general question of biogenesis from abiotic material. The origin of life as an ETI refers specifically to the transition from the very first individuals (the issue of what they were is discussed below) capable of reproduction to another kind of individual that was more stable, was viable, and demonstrated all the properties associated with a living system (see Penny's criteria of life in chapter 1). In a sense this is analogous with what Gánti called *potential* and *absolute* life criteria (chapter 3). The individuals at the lower level in the ETI are more closely aligned with potential life criteria, although they do demonstrate some of the features associated with life, while the individuals at the higher level

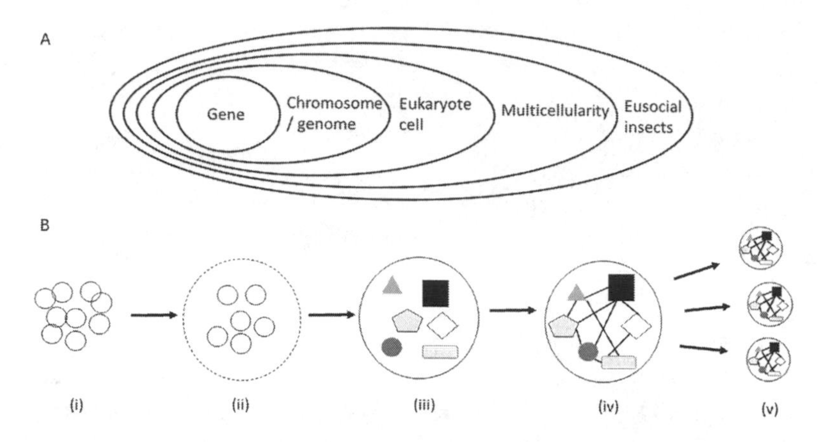

FIG. 7. Evolutionary transitions in individuality. ETI theory explains how individuals cooperate at one level to form a new kind of individual. This happened several times during the history of life. (A) Genes (see chapter 13 for my interpretation of a gene) cooperated and eventually became functionally integrated to form chromosomes or genomes (or cells if one prefers the cell biology perspective). Prokaryote cells (which explain the presence of nuclear, mitochondrial, and plastid genomes in eukaryote cells) cooperated in eukaryogenesis. Multicellular organisms formed from unicellular ancestors and eusocial insects like the hymenoptera exist in specialized communities. (B) In the first step of an ETI (i) individuals of a particular kind (ii) form loosely associated groups, colonies, or aggregates; (iii) the groups become more defined, discrete entities and following cycles of cooperation, conflict, and conflict mediation, individuals within the group become functionally specialized (depicted by different shapes) and lose their individuality such that (iv) a new kind of individual emerges from the functional integration of its components. This new kind of individual is also a target for natural selection and (v) gives rise to offspring of its own kind. (These are my interpretations of the steps involved, but see also Michod 2007 and Hanschen et al. 2015.)

exhibit, in a stable and viable way, all the properties Gánti referred to as absolute life criteria. There is a large body of work examining the origin of life as an ET. As an ETI, the origin of life is less well explored (but see Agren 2014; Durand and Michod 2010; Koonin 2016; Szathmáry and Maynard Smith 1997; Michod 1983), although elements of the theory of ETIs such as conflict and cooperation, group selection, and the units and levels of selection (see chapter 3) have been incorporated into some of the theoretical works. This is because, as discussed earlier, there are no extant molecules that are sufficiently representative of the first replicators—the individuals at the lower level.

An interpretation of the origin of life as an ETI is valid in its own right (Agren 2014; Durand and Michod 2010; Koonin and Martin 2005; Koonin, Wolf, and Puigbo 2009). It is difficult to see how life began without integrating a transition from lower-level replicators to the higher-level ones described in this chapter (see also the synthesis in chapter 8). There are inherent biochemical limitations in the earliest replicating units, and replicators that increase their biochemical complexity are easily outcompeted by those that remain less complex (see chapters 2 and 3). The functional repertoires of the earliest replicators were highly constrained. Overcoming this constraint relied on cooperation and the functional integration of groups of replicators. Incorporating such a transition using the theory concerning ETIs into explanations of the earliest living systems provides a more plausible synthesis.

ETI theory is especially useful for examining life's beginnings because it reconciles the disconnected philosophical, theoretical and biochemical points of view (chapters 3–6). The argument that the origin of life was not a discrete event is inherent in the theory of ETIs, which conceptualizes the process rather than the abiotic-biotic boundary (Radzvilavicius and Blackstone 2018). This boundary is an artificial construct; the origin of life was not a binary, before-after, phenomenon. ETIs cannot be reduced to a single event and capture the process of life's origins. The theory of an ETI also encourages us to think concretely about what exactly the individuals (replicators) were at the lower and higher levels. As alluded to earlier, the theoretical work on replicators has often been abstract and sometimes divorced from the mechanistic constraints discovered in the experimental works. By interpreting the origin of life as an ETI "from the bottom up" (Radzvilavicius and Blackstone 2018) we are forced to base our concept of replicators on specific biochemical knowledge, on what the units of replication may have been, and on how the transfer of fitness from the lower to the higher level may have occurred.

The hypothetical replicating units at the lower and higher levels

The most parsimonious scenario is that the first replicating unit did not comprise DNA or proteins greater than a few amino acids. The RNA world hypothesis (chapters 3–5) invokes a putative primordial living system that comprised RNA molecules with enzymatic potential. It also seems impossible that the first replicating unit was a single molecule (chapters 3, 5 and 6). Rather, the very first entity capable of self-replication must have comprised a network (chapters 2 and 3). The network of interacting molecules enabled replication of the collective (see Lincoln and Joyce 2009 for a system of two ribozymes and four oligonucleotide substrates). The theoretical, evolutionary, and empirical data discussed in previous chapters, are all drawn upon in this chapter and the next, and to avoid confusion it is prudent to highlight, again, the different terminologies used by biochemists and evolutionists (see also chapter 15, especially additional notes 15.1). When biochemists refer to groups of molecules as being replicators, the word *group* has no special significance except that it is a collection of inanimate molecules. The molecules replicate because of their ability to interact with each other—the individual molecules in the group are not autonomous in any way. To evolutionists, the term *group* usually has a group selection context, and refers to a group of individuals, but each individual has the potential to reproduce autonomously. The difference between the biochemical group and the evolutionary group will be pointed out where necessary in the subsequent discussion, but this is an important distinction. Ignoring this frequently leads to misunderstandings. Biochemists may not appreciate the loaded nature of the term *group* used by evolutionists. At the same time, evolutionists end up confused when biochemists refer to the groups when they are simply referring to collections of inanimate molecules. At first glance the difference may seem obvious and even unnecessary to mention. However, researchers in the different disciplines often speak at cross purposes at conferences and scientific meetings. Mechanistic biologists (cell biologists, biochemists, physiologists, etc.) and evolutionists ask very different fundamental questions and have very different approaches to explaining phenomena.

The terms lower-level replicating units (LRUs) and higher-level replicating units (HRUs) are used to refer to the two levels in the ETI at the origin of life. These units existed prior to the emergence of protocells. Each LRU comprises a group of molecules (in the biochemical sense) where the entire group replicates because of the interactions between molecules. There is no single molecule that is a "replicator" at the lower level. A population or group (in the evolutionary sense) of LRUs, however, comprises individuals that may vary in their fitness because of random mutations and genetic drift. Lewontin's criteria (chapter 2) of heritability, variation, and fitness in a reproducing population are met (Lewontin 1970) and the population would have evolved by natural selection. ETI theory explains the process by which HRUs emerged from populations of LRUs via group selection (the term *group* is used again in the traditional evolutionary sense in this instance).

We have a general idea of what the LRUs were—they comprised a biochemical collection of ribozymes and ribonucleotide polymers (such as the system in Lincoln and Joyce 2009). What can we use as an exemplar of an HRU? Unfortunately, there is no completely satisfactory answer to this. Daly and colleagues have pointed out (Daly, Chen, and Penny 2011) that RNA almost certainly preceded DNA, but a purely RNA-based HRU that emerged from LRUs has not been developed experimentally. Ideally, we would want to use (biochemical) collections of extant and engineered ribozymes as the functional analog of an LRU and (evolutionary) groups of LRUs to develop HRUs. However, manipulating LRUs experimentally to evolve into an HRU would be very challenging indeed. It would certainly be a remarkable technical accomplishment that would reveal important biochemical features concerning ribozyme cooperation and conflict. But it should also be remembered that even if this was achieved, it is unlikely that it would be wholly accurate. The reason is that by the time ribozymes had emerged, amino acids, small peptides, and possibly many other cofactors coevolved with the LRUs before the transition to the higher level (van der Gulik and Speijer 2015) (ribosomes and ribonucleoproteins are extant examples of RNA-protein coevolution). It is unlikely we will know how these molecules interacted or even which molecules played which roles except that LRUs and their constituent ribozymes are important for tracing the transition from LRUs to HRUs.

Could the hypothetical first gene and genome (Juhas, Eberl, and Glass 2011; Koonin 2003) be used as exemplars of LRUs and HRUs, respectively? On the one hand this is also unrealistic because, as mentioned above, the genes and genomes, as we understand them today, make use of a genetic code and large proteins for them to be functional (even if the hereditary material of some classes of selfish genes do start out as RNA, they still require complex DNA and protein steps). The first LRUs and HRUs would have comprised RNA with amino acids and peptides functioning as cofactors, but this would have occurred before a fully fledged genetic code for protein synthesis existed. On the other hand, the gene-genome divide is very helpful in that it is a useful model for investigating the sociobiology of selfish replicating units (like the mobile genetic elements or MGEs discussed later). The ribozyme model systems are limited in this aspect, in that they have not been developed as HRUs. MGEs do, however, reveal much about the conflict, cooperation, and functional integration between self-replicating molecules at the two levels (Agren 2014; Durand and Michod 2010; Kidwell and Lisch 2000; Koonin 2016).

MGEs are autonomous (to varying degrees) and as such are sometimes considered functionally analogous to the LRUs. They are also well documented as drivers of genome evolution (Kazazian 2004; Malik, Burke, and Eickbush 1999). Over time they may lose their autonomy, become functionally integrated in the genome, and carry out specialized tasks at the higher level (Brosius 1999). MGEs and the role they play in the gene-genome transition are valuable tools for investigating the sociobiological aspects of the origin of life. But, as discussed above, there are limitations because the highly sophisticated genetic code and protein translation did not exist at the very beginning of life. The insights gained from MGE model systems, therefore, are only a proxy for what may have happened at the very beginning, before proteins, as we understand them today, existed (Durand and Michod 2011).

In the absence of an ideal model system, elements of both ribozymes and MGEs are used as exemplars to provide a hypothetical recapitulation of the origin of life as an ETI.

The transition from LRUs to HRUs

Briefly, the hypothesis of the origin of life as an ETI is as follows. The first LRUs were relatively simple networks of at least two ribozymes (and their substrate oligonucleotides) each capable of replicating the other. The network reproduces as a single individual LRU and is considered a group in the biochemical sense. This is because the network comprises ribozymes that copy each other, which is an example of enforced molecular reciprocity (see chapter 13 for a more detailed discussion of reciprocity). A laboratory model for such a minimal LRU system has been developed (Lincoln and Joyce 2009). A population of LRUs would have comprised competing individual LRUs, each investing solely in its own fitness, which can be described in terms of viability and reproduction. Groups (in the evolutionary sense) of LRUs formed such that, assuming the required conditions existed as predicted by the theoretical, geochemical, and biochemical data (chapters 3–5), the potential for between-group (as well as within-group) competition arose. ETI theory explains how competition between individuals within the group decreases group fitness, while cooperation between LRUs allows for viability or reproduction of the whole group to increase. When the variance between groups is greater than the variance within the group, that is, between LRUs in the group, group selection can overcome selection between LRUs. Groups with greater cooperation between LRUs outcompete those with fewer cooperators and as the evolutionary interests of LRUs within cooperative groups become more aligned, individual LRUs no longer invest solely in their own fitness. They are free to evolve new functions that enhance group fitness further, either viability or reproduction components. MGEs have been used as a model system for tracking this transition (Agren 2014, Brosius 1999, Durand and Michod 2010, Kidwell and Lisch 2000, Koonin 2016, Malik, Burke, and Eickbush 1999, Sinzelle, Izsvak, and Ivics 2009). After cycles of cooperation, conflict, and conflict resolution, LRUs eventually relinquish their own individuality and invest completely in the group, the emerging HRU. The HRU becomes ever more functionally integrated, eventually emerging as a new kind of individual, where LRUs are completely invested in the HRU and no longer recognizable as discrete individuals. The HRU is an indivisible individual with functional divergence in its component parts and more complex (however we wish to define complexity [Adami 2002]) than the LRUs from which it arose.

The empirical RNA-world data suggest that the most likely mechanism for ribozyme-mediated replication in the LRUs was either ligation, polymerization, recombination, or a combination of these (chapter 5). No other functions are necessary for LRUs to reproduce. The hypothesis is that over time, as the group of LRUs became more integrated, some of the LRUs were free to specialize in additional functions. Some may have specialized in replication of the group (as opposed to individuals within the group); others could have evolved new functions that increased group viability, such as the biosynthesis of mononucleotides and oligonucleotides that were used as substrates by those ribozymes responsible for replication. In addition, ribozymes that promote the synthesis of amino acids and other molecular cofactors, which coevolved with RNA (van der Gulik and Speijer 2015) and enhanced group-level fitness, would have emerged. LRUs, however, were severely constrained in terms of function and the amounts of hereditary information they contained. Their cooperation was required for complexity to increase. One of the features of HRUs is the evolution of policing mechanisms that control the temptation of LRUs to revert to selfishness. This may have included cleavases, ribozymes that can cleave non-cooperators that make the group vulnerable to invasion by selfish LRUs.

HRUs comprising ribozymes with divergent functions have not been observed in nature (certainly not in the way that they are described here), nor have experimental biochemists reported attempts to evolve such a system. LRUs have been obtained experimentally, such as those developed by Lincoln and Joyce (Lincoln and Joyce 2009) and Hayden, Lehman, and colleagues (Hayden and Lehman 2006; Higgs and Lehman 2015). The scenario described here is therefore conceptual, based on theoretical data and supported by the model experimental systems available to us. It is unlikely we will know exactly how this transition occurred. We can be certain, however, that something like this conceptual LRU-HRU ETI must have happened for a large, sophisticated chemical system capable of reproduction to exist. It represents, perhaps, the most parsimonious explanation. Importantly, it is rooted in our knowledge of early life biochemistry and consistent with the conceptual, theoretical, and experimental work concerning the gene-genome ETI (Agren 2014), and provides a useful heuristic guide for exploring the sociobiology of the earliest living molecules. Without an account of the LRU-HRU transition described above (à la Agren 2014, Durand and Michod 2010, Michod 1999, 2007, Michod and Nedelcu 2003), a more complex molecular system with greater genetic potential was not possible and cellular life could not have emerged. HRUs may have existed well before the origins of genes and genomes or the HRU may have coevolved with DNA, the genetic code, and proteins and been a kind of proto-genome. There is no way of knowing the timeline of events for certain, but whatever the stages, the evolution of an HRU would have been necessary for genomes as we know them today to emerge.

Extant genomes still bear the hallmarks of a gene-genome ETI and we can extrapolate these features to the conceptualization of the LRU-HRU ETI in the RNA world. It is clear that "genomes were forged by massive bombardments with retroelements and retrosequences" (Brosius 1999). Retroelements and other types of MGEs (sometimes also called transposable elements) still comprise significant proportions of genomes (see additional notes 6.2). From an ETI perspective, one of the compelling discoveries was that MGEs and viruses have had a major impact on the formation of genomes (Daugherty and Malik 2012). As Brosius (Brosius 1999), Kidwell (Kidwell and Lisch 2000), Sinzelle (Sinzelle, Izsvak, and Ivics 2009), and others have discovered, MGEs start life entirely selfish (investing only in their own fitness) but over time may become domesticated by the host genome, behave altruistically, couple their evolutionary fate with that of the genome, and eventually lose their individuality. As Kidwell says, "At least two levels of selection can apply to transposable elements" (Kidwell and Lisch 2000), and Koonin makes the point that they are central to ETIs (Koonin 2016).

Aside from the two levels of selection and the two types of replicating units at each of the levels, there is the important question of fitness transfer (although it must be said that for some philosophers the export-of-fitness view is simply an epistemic exercise and moot, see Bourrat 2015). How exactly does an LRU give up its fitness and invest in another kind of individual (the HRU)? This is a compelling question asked of any ETI and has been addressed mostly in the context of the unicellular-multicellular ETI (Michod 2006, 2007; Michod and Nedelcu 2003; Michod 1999). All else being equal, natural selection does not favor altruists in the population. To appreciate the issue of fitness transfer at the origin of life, we again turn to the gene-genome model system to guide us through the mechanics.

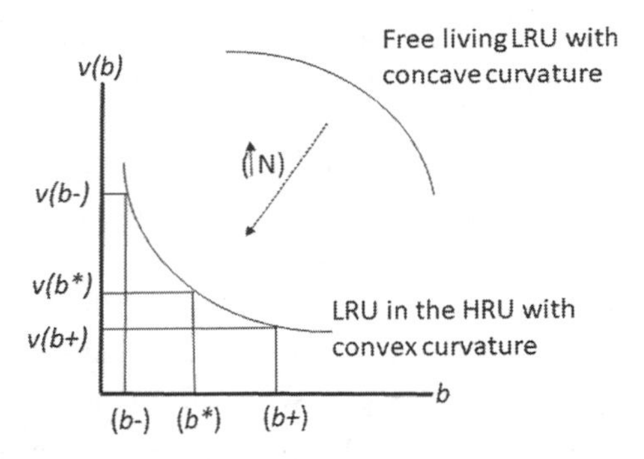

FIG. 8. Life-history trade-offs in LRUs and the augmentation of group fitness. The hypothetical trade-off relations between viability / survivability (v) and fecundity / reproduction (b) for an LRU when free living and as part of the HRU are shown. (A) A trade-off with concave curvature is predicted for the free-living form where selection acts on the LRU. (B) A change in the v–b trade-off to a convex curvature augments group fitness and as group size increases (↑N) this augmentation increases further. The progression to convexity results in the specialization of the two fitness components by members of the group. The advantage of specialization is evident from the finding that $v(b-) + v(b+) > 2v(b^*)$ for points on the convex curve, where b− and b+ are two points equidistant from any point b*. The image is modified and redrawn from Michod 2006 (with additional interpretation appropriate for the question at hand).

How was fitness transferred from LRUs to HRUs?

This issue can be explained by what Michod calls the *covariance effect* (Michod 2006). The covariance effect was primarily used to explain functional specialization in the two fitness components of viability and reproduction in multicellularity evolution, but it can be extended to the question of fitness transfer. Covariance is a measure of how two parameters vary in relation to each other and for ETIs, the covariance between the two most fundamental life-history fitness components (reproduction and viability) has an unavoidable effect on fitness transfer (see additional notes 6.3). One of the first emergent properties found in groups of replicating units is the group-level covariance between reproduction and viability. Covariance in this case is negative, which means that the two components trade off against each other. An investment in one component diminishes investment in the other. The nature of this trade-off can vary and by evolving a life-history strategy that lowers one fitness component in favor of another, fitness at the higher level is increased when there is convexity in the trade-off (fig. 8). Using MGEs as the model system, MGEs can enhance genome-level fitness by modifying their life-history strategy (see the trade-off curves in fig. 8 and see Michod 2006 and references therein for a detailed explanation). The endgame is that the fitness of the group is augmented over the average fitness of individuals within the group by the covariance effect and the greater the number of individuals in the group the greater the augmentation of group-level fitness (for a specific example see fig. 2 in Durand and Michod 2010). (The aim here is to provide a very brief justification for how the problem of fitness transfer, which is so central to ETIs, is overcome. Clearly, this is a complex question, and for a full appreciation of what is meant by the covariance effect and the associated mathematical equations, a much more detailed discussion of the theory concerning life history evolution is required. This is beyond the scope of this book and the reader is referred to the references in the additional notes 6.3). This emergent phenomenon of group-level covariance and the optimization of group fitness explains how the change in a life-history trade-off leads to functional specialization (referred to as *division of labor* but, see additional notes 6.4) of LRUs and the evolution of a new kind of individual, the HRU.

The origin of life as an ETI brings together some of the disparate information from philosophy, theoretical modeling, evolutionary theory, ecology, and biochemistry into a single conceptual framework. This framework, in conjunction with the discussions in previous chapters, is necessary to cover all the major steps in a broad synthesis for the origin of life.

7
A synthesis for the origin of life

The philosophical, theoretical, genomic, and empirical data discussed in the previous chapters can be integrated to develop a broad synthesis for the origin and evolution of the first forms of life. It is unlikely that an exact recapitulation of how and why living forms arose is ever going to be possible, but based on the integration of current knowledge, a parsimonious scenario can be sketched out. Not having an entirely agreeable definition for *life* that satisfies all scientists (chapters 1 and 2) means that there is not only uncertainty about the process, but also considerable disagreement. Nevertheless, a step-by-step account of the most important evolutionary innovations and an account that incorporates the essential components from the different disciplines, which most syntheses of the origin of life do not do, is possible. There is also likely to be considerable overlap between steps. They are certainly not as distinct as portrayed here. The endpoint of this synthesis, however, is to arrive at an entity (an "individual" in the evolutionary sense) that has all the key properties that differentiate abiota from biota (chapter 2). This entity is abstract in the sense that it is not known what it was, but it is also realistic and based on the biochemical, theoretical, genomic, and computational data. It is not known what these first "individuals" were, but for evolutionists it is important to arrive at a population of individuals on which natural selection can act. Non-adaptive processes and stochasticity in the system is largely ignored in the proposed synthesis. This is, of course, a major caveat, but at the same time evolution by natural selection must have played a major role. Living systems cannot evolve and be maintained by non-adaptive or stochastic processes alone. In fact, that is one of the principles of biology discussed in chapter 2. The aim here is to understand how life was fashioned via natural selection based on the knowledge currently available. At the same time, for mechanistic biologists, an account of the first living individuals must be realistic and rooted in empiricism (including the genomic and theoretical data). This synthesis, however, while remaining true to the experimental data focuses primarily on the evolutionary aspects. The metabolic components, for example, were clearly essential for the maintenance of early life but they are not included in this evolutionary-based synthesis except where they are relevant for the evolutionary process. This account is also consistent with an RNA world in which RNA emerged first and coevolved with cofactors, amino acids, and small peptides. DNA was the last of the information molecules to evolve, possibly at the time when the genetic code and the simplest protocells first emerged.

Many biologists tend to trace life back to the earliest protocells (a LUCA-like organism, discussed in chapter 3). But even the earliest cell-like structures were extraordinarily complex, and a considerable amount of evolution happened before they emerged. To appreciate how life started, it is essential to understand the living systems long before cells arose. From an evolutionist's perspective, an explanation is sought that will bridge the gaps from the first biologically relevant molecules to the earliest protocells.

The evolution of life in eight steps

STEP 1: THE EMERGENCE OF BIOLOGICALLY RELEVANT MOLECULES (FIG. 9)

An appreciation of the biochemistry and geochemistry of the early Earth has allowed chemists to provide an outline for the emergence of the nucleobases (nitrogenous compounds), nucleosides (nucleobases with sugar moieties), and nucleotides (nucleosides with phosphate groups). Simple amino acids and cofactors may have arisen at the same time. Despite the advances in our understanding of the early chemistry, there are still significant gaps in explaining how this happened. In particular, the issue of the reducing atmosphere of the early Earth is a sticking point and the origin of some of the precursors (sugars, inorganic phosphate, etc.) of the biologically relevant molecules is disputed (chapter 4). These unanswered questions, however, are not insurmountable obstacles to a synthesis of life's origins. All the important chemical reactions may not be dissected, and some precursors may have been deposited on Earth by extra-terrestrial bodies like meteorites, but in one way or another, the biological building blocks emerged by geochemical means. The important conclusion at this step is that these early molecules, or something like them, arose. None of these can be considered living in any meaningful sense. The processes occurring at this stage were purely physical and chemical in nature.

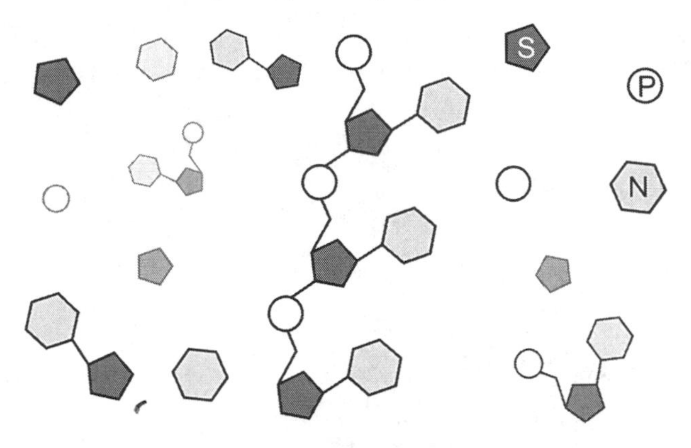

FIG. 9. The emergence of biologically relevant molecules. Nitrogenous bases (N), ribose sugars (S), and inorganic phosphates (P) emerged on the early Earth, giving rise to the ribonucleotide monomers that formed the basis for the RNA world.

STEP 2: PASSIVE REPLICATORS (FIG. 10)

Given the appropriate chemical conditions, ribonucleotides formed polymers spontaneously, particularly when in contact with materials like montmorillonite clay, which facilitates the polymerization reaction. Once formed, single-stranded polymers (ssRNAs) had the potential to form complementary base pairs with other ribonucleotide monomers or polymers, such that double-stranded ribonucleotide polymers (dsRNAs) were produced. Molecules like these were able to reproduce themselves passively by strand dissociation and renewed base-pairing with monomers or oligomers present in the environment, resulting in newly formed double-stranded molecules. The replication of ssRNAs in this fashion, although requiring chemical energy, was still passive in the sense that there were no enzymes to catalyze the reactions. A crucial limitation was that the error rate via this form of replication was prohibitive. It was far too high for copies of individual replicators to be reproduced faithfully (chapter 3). The system was in constant flux with new ssRNAs forming, decaying, and passively copying themselves inaccurately. The poor fidelity meant that the next generation of replicators were poor copies of their parents and the hereditary nature of the system was chaotic.

STEP 3: ENZYME-MEDIATED REPLICATION (FIG. 11)

For the earliest forms of life to evolve by natural selection, the fidelity of replication from parents to their offspring was important. This could be

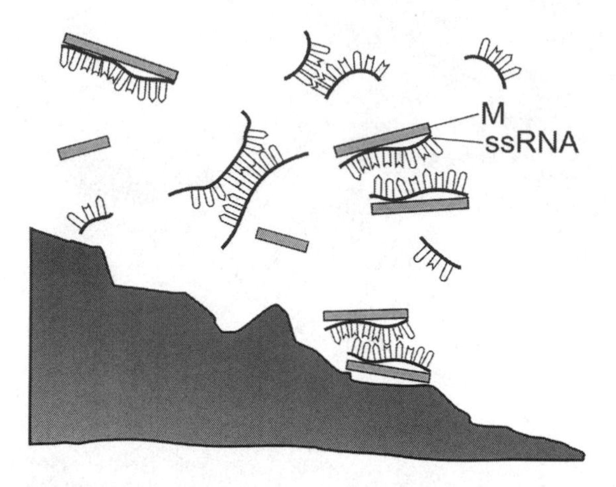

FIG. 10. Passive replicators. Clay matrices like montmorillonite (M) facilitated the polymerization of short ssRNAs that were able to replicate passively by error-prone complementary base-pairing.

achieved only by enzyme-mediated replication so that the nucleotide sequence of the parent molecule was faithfully reproduced. The problem, however, was that enzyme activity is found only in longer ribonucleotide polymers (ribozymes) and the question arises: how was a population of large polymers maintained if they were not copied accurately? There are several potential routes around this problem. The theoretical data suggest this may have been via clouds of quasispecies or hypercycles (chapter 3). The empirical data indicate that recombination and ligation events seem to be the most likely mechanisms by which polymers with polymerization activity (the enzyme reaction required to copy a strand of RNA faithfully) emerged (chapter 5). One of the solutions lies in the finding that a collection of small RNA polymers can function like a much larger polymer with recombinase activity. An alternative is that small polymers can have ligase activity. Small RNAs, therefore, could have joined other RNA strands to themselves such that one of the new, larger polymers had polymerase activity (a chance event). These two potential routes have been demonstrated experimentally. Some of the groups of molecules (LRUs in chapter 6; see also the discussion concerning the distinction between biochemical and evolutionary groups) that have been shown to be self-sustaining have been the collections of simple ligases with their substrate oligonucleotides (Lincoln and Joyce 2009), the recombinases with their substrates (Hayden and Lehman 2006), and simple ligases with a pool of random oligonucleotides (Dhar et al. 2017).

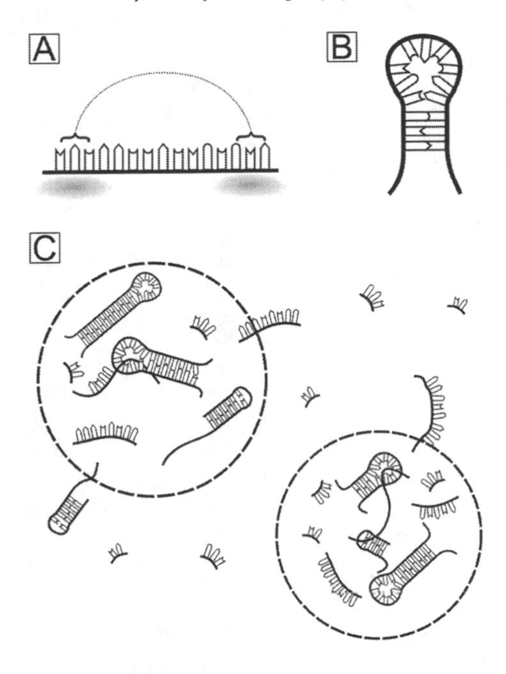

FIG. 11. Enzyme-mediated replication. (A) Complementary base-pairing of short polymers allowed (B) ribozymes to emerge from their secondary and tertiary structures. (C) Collections of recombinases, ligases, and possibly other short-polymer ribozymes, as well as oligonucleotide substrates, formed the precursors of LRUs that were eventually able to copy the collection of abiotic molecules in a compartment. For practicality, the theoretical data that invoke clouds of quasispecies or hypercycles (chapter 3) are not included in this step because there are very few model systems that can be used as exemplars of these molecules.

STEP 4: COMPARTMENTALIZATION (FIG. 12)

Compartmentalization was required in some form or another, so that the molecular components of the replicating system remained in close contact with each other and for selection to occur on the molecular collectives in steps 5 and 6 below. Each compartment was a collection of ribozymes, their substrates or oligonucleotides, and cofactors or amino acids. Compartmentalization was also essential to protect hypercycles or groups from invasion by parasitic molecules (selfish, non-cooperators), which could infiltrate the collections of molecules, causing their interactions to falter. The empirical findings suggest that compartmentalization may have taken the form of lipid micelles, iron casings, or water droplets such that offspring emerged from their parents. Alternatively, the coevolution of molecules may have led to physical associations without the need for

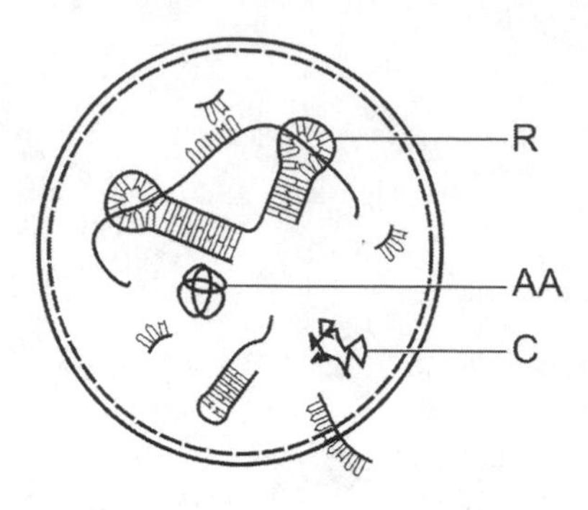

FIG. 12. Compartmentalization. Compartmentalization was an essential step for the functionality of the collections of ribozymes (R), RNA oligomers, and other possible cofactors (C) or amino acids (AAs) that may have coevolved with the RNAs. Compartments, which may have been water droplets, lipid micelles, or iron casings, provided molecular proximity and protected against invasion by molecular parasites. The coevolution of interacting molecules may also have led to physical associations.

structured compartments. Although compartmentalization provided a primitive structure in which the biochemical reactions were carried out, there was still a degree of fluidity with ribozymes, their substrates, and cofactors migrating between them (see for example the SCM, stochastic corrector model, in chapter 3).

STEP 5: LOWER-LEVEL REPLICATING UNITS (LRUS) (FIG. 13)

Each compartmentalized group of molecules (the LRUs or lower-level replicating units in chapter 6) that existed at this stage constituted a single replicating system. The term *group* here is used in the biochemical sense, not the evolutionary sense (chapter 6 and additional notes 7.1), in that it is a collection of non-autonomous molecules that interact with each other. None of the molecules in the collective can replicate themselves individually. They can, however, exhibit reciprocal molecular interaction by copying parts or all of each other, or benefiting each other by ligation reactions. Each LRU, however, is an autonomous individual capable of independent reproduction (chapters 3, 5, and 6). An example of an LRU may be a simple two-membered hypercycle or the collection of two ligases and their four substrates described by Lincoln and Joyce (Lincoln and Joyce 2009). It is the LRU that is a replicating individual and populations of LRUs were subject to evolution by natural selection.

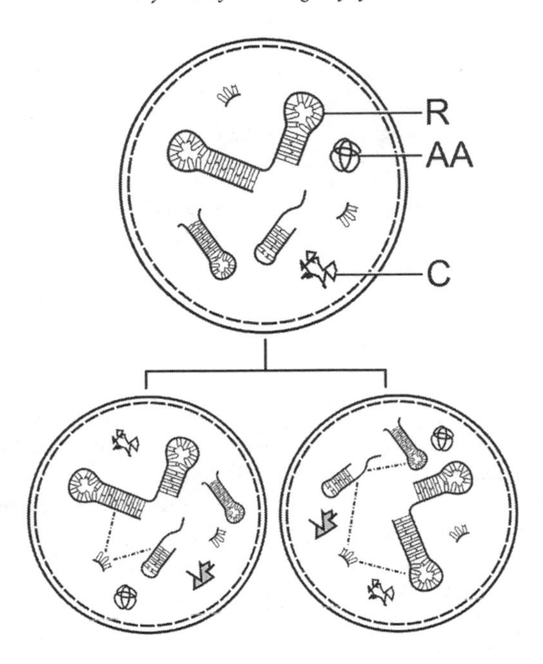

FIG. 13. Lower-level replicating units (LRUs). Collections of ribozymes (R), RNA oligomers, cofactors (C), and amino acids (AA) functioned as LRUs. LRUs replicated, with daughter LRUs exhibiting phenotypic variation. For example, the dashed lines show differences between structural and functional connections between molecules in daughter LRUs.

STEP 6: FUNCTIONAL AND STRUCTURAL INTEGRATION OF THE COMPONENTS MAKING UP LRUS (FIG. 14)

When LRUs first formed, their molecular components were only loosely connected and reciprocal and mutual dependencies were likely weak or minimal in number. Over time, selection drove the evolution of LRUs that were more integrated, allowing for their molecular components to diversify. For example, some molecules may have been responsible for replicating all the components of the LRU. Others may have diversified to become synthases of ribonucleotides, metabolites, or components of the material making up the outer structure or membrane. The endpoint of this step is that there was a population of LRUs where each LRU consisted of a collection of molecules. LRUs exhibited varying degrees of structural and functional integration.

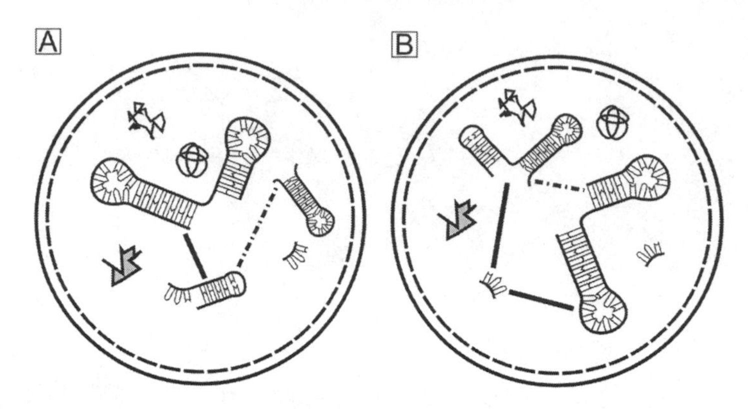

FIG. 14. Mutual dependencies of the molecular components in LRUs. LRUs with greater functional and structural integration of the component molecules are selected for. For example, (A) may outcompete (B) due to the differences in component molecules and produce more offspring. Functional diversification of molecular components followed. For example, where there were two ribozymes performing similar functions, one is free to evolve new functions.

STEP 7: THE EVOLUTIONARY TRANSITION TO HIGHER-LEVEL REPLICATING UNITS (HRUS) (FIG. 15)

Despite the evolution of more stable LRUs, they were still functionally constrained because of the limited information in the ribozymes. Cooperation between LRUs resulted in mutual dependencies or reciprocal altruism developing between individual LRUs. Cooperative groups (the term *group* is used here in the evolutionary sense, not the biochemical sense, chapter 6) may have formed and over time group selection came into play. The evolutionary transition in individuality discussed in chapter 6 explains how HRUs evolved from LRUs after cycles of conflict, conflict mediation, and cooperation. The important features of any ETI, as described by Buss (Buss 1987), Michod (Michod 1999), Ruiz-Mirazo and Moreno (Ruiz-Mirazo and Moreno 2012), and many others, like the division of labor and transfer of fitness are eventually accomplished such that a new "kind" of individual emerges.

STEP 8: PROTOCELLS (FIG. 16)

It is unlikely that there was a single, individual HRU that can be considered the first protocell. What is much more plausible is that the first cells emerged from HRUs that were much less structurally discrete than the

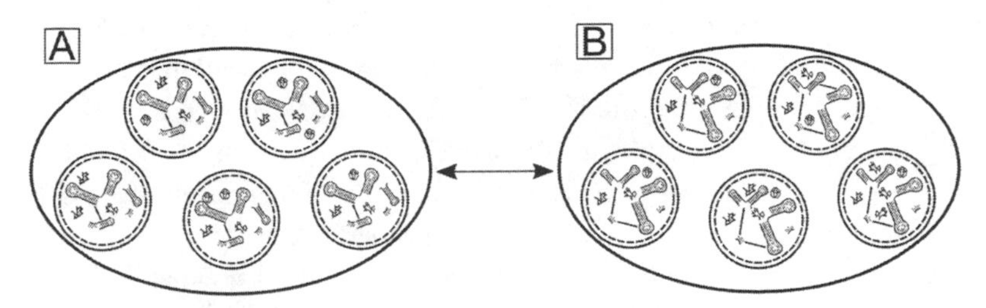

FIG. 15. Higher-level replicating units (HRUs). Groups (A) and (B) may compete against each other. When the conditions are right and the variance between groups is greater than the variance within groups, within-group competition becomes secondary to between-group competition. An ETI from LRUs to HRUs is completed after rounds of conflict, conflict mediation, and cooperation.

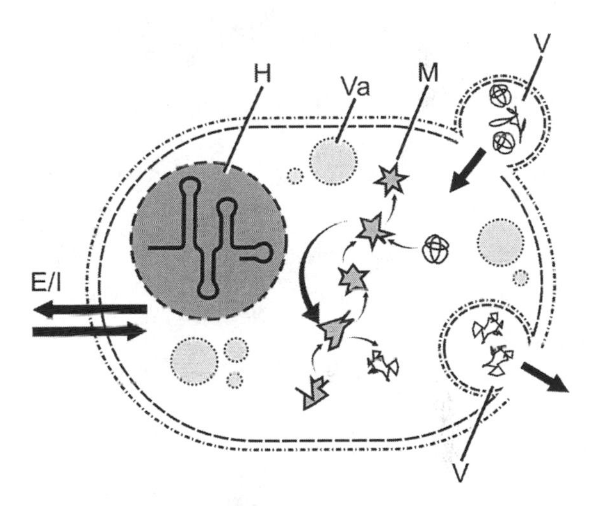

FIG. 16. The protocell. Protocells emerged from HRUs. They contain hereditary (H) components (ribozymes, RNAs, simple amino acids, and cofactors, a primitive genetic code) and a primitive metabolism (M). There is efflux (E) and influx (I) of simple compounds (water, salts, etc.). Vesicles (V) may enter or leave the protocell and vacuoles (Va) are intracellular stores of metabolites, salts, water, etc. Note how similar this is to Gánti's chemoton discussed in chapter 3.

cells that occur today. The cellular borders may have been more amorphous, sometimes fusing or budding off and changing in their chemical composition. This is in keeping with the philosophical, theoretical, genomic, and empirical data discussed in earlier chapters, all of which converge on the idea that there was no singular "LUCA." Rather, there was extensive lateral exchange of hereditary material between HRUs. Our understanding of the first protocell is still more abstract than concrete and

Gánti's chemoton remains a useful way for thinking about these early cells. The important step is that, at this stage, there is a protocell that contains some form of hereditary material (ribozymes and a primitive version of the genetic code), a simple metabolism for maintaining the cell's viability (amino acids, small peptides, and cofactors), and a mechanism for maintaining structural integrity (a lipid membrane with proteinaceous components). All the components or properties of life that are demanded by the different disciplines are present. From the protocells emerged the two prokaryote domains, the archaea and the eubacteria.

PART TWO

The origin of death

Die not, poore death, nor yet canst thou kill me.
From rest and sleepe, which but thy pictures bee,
Much pleasure, then from thee much more must flow.

JOHN DONNE, "Sonnet X"

8
Philosophical considerations and the origin of death

The issue of mortality has occupied an enduring place in the history of human thought. Epicurus (341–270 BC) was one of the early Greek thinkers who considered the meaning of death and its place in the natural world, even if he was somewhat reluctant to engage fully with the topic. In his letter to Menoeceus he offered the view that "death . . . , the most awful of evils, is nothing to us, seeing that, when we are, death is not come, and, when death is come, we are not" (see Luper's discussion of Epicurus' rejection of the Harm Thesis in Luper 2016). Death "is nothing to us," he suggested, and perhaps hardly worth thinking about, since after death the individual ceases to exist and before death, death has not occurred and is harmless. Central to the Epicurean view is the claim that an individual's soul or mind is not immaterial. It cannot be separated from the body and does not exist after death. In one sense, Epicurus taps into the thinking of Thales and Anaximander (see chapter 1), who separated rational thought and observation from theology. Both wished to examine the meaning of life in a rational way without theological influences. Similarly, Epicurus separated the natural world, including death, from the world of the gods, arguing that after death nothing in the natural world continued in the world of the gods. The work of Epicurus is covered in detail by Luper, who also gives a modern and comprehensive treatment of the philosophy of death (Luper 2009).

Philosophical interpretations of mortality, from ancient times to the present day, have usually concerned death in multicellular organisms (typically plants and animals, including humans). Questions concerning the

I am very grateful to Grant Ramsey for his reading, corrections, and comments on this chapter. At the same time, any errors are mine alone.

death of individual cells also began in the context of multicellular organisms. Cell death (the different forms were not known at the time) was first observed as a part of normal development in multicellular embryonic tissues by Collin (Collin 1906), Ernst (Ernst 1926 and references to Muhlmann therein), Hamburger and Levi-Montalcini (Hamburger and Levi-Montalcini 1949), and Kallius (Kallius 1931). The specific terminology of programmed cell death (PCD) was introduced decades later and made explicit by Glücksmann (Glücksmann 1951) and Lockshin (Lockshin and Williams 1964), arising chiefly from studies of normal animal ontogeny.

Mortality in single-celled organisms was, at first, not given serious consideration. It was assumed they died via extrinsic causes like starvation, predation, or physical and chemical damage, or they simply accumulated deleterious mutations as they aged and lost viability over time. It has emerged, however, that unicellular organisms are not immortal and non-incidental forms of microbial death existed long before multicellular life evolved. In the last few decades of the twentieth century, researchers began actively investigating whether a process like PCD could occur in microbial life. The subject of unicellular mortality has subsequently become part of mainstream scientific endeavors. What began as a minor curiosity is now an area of in-depth research and is the focus of part 2. While there is an extensive body of work on the evolution of death in multicellular organisms, this topic is not covered in this book except for the section on behavioral suicide in insects and arachnids in chapter 16. Instead, the focus is the often misunderstood and overlooked subject of non-incidental death in single-celled organisms.

The concept of programmed cell death

The very concept of PCD in microbes is challenging to articulate. This is true in part, because it is important to appreciate what is meant by death generally. Defining *death*, as with the definitions of life in chapter 1, is not a trivial matter. Superficially, one might define death as simply the extinguishing of life. But, when looking more closely at what was meant by life in part 1, the question of death becomes more complicated. Some researchers understand life in terms of the properties associated with it. Penny listed these as being an energy source (and energy gradient), basic biochemical reactions driven by the energy gradients, organization (membranes, compartmentalization, and separation from the external environment), and self-reproducibility (genetic heritability, information transfer, evolvability, etc.) (Penny 2005). In one sense, therefore, death is the loss of these properties in an entity that once had them. Other researchers take a systems biology–centered approach, like Joyce and others who define life as "a self-sustaining chemical system with the capacity for Darwinian evolution" (Joyce 1994). In this sense, then, death is the loss of the sustainability of the system and the subsequent loss of capacity for Darwinian evolution. However, as discussed in chapter 1, the definitions of life themselves are problematic. None of the definitions succeed in capturing its essence, and different researchers tailor the definition to make it fit a particular research question (Casiraghi et al. 2016). Saying that death is the absence of life is, therefore, hardly satisfactory. Without a general definition of life, defining death in terms of the cessation of life is not helpful, even if the two are intuitively oppositional. Contrasting death with Penny's property-based definition of life, for example, provides little in the way of explaining the nature or evolution of death. Similarly, Joyce's interpretation of life, cannot serve as a basis for the definition of death. It provides no information about why or how the loss of self-sustainability or the capacity for Darwinian evolution occurs. There is also, of course, the issue of evolution by non-Darwinian means, which is not included in Joyce's interpretation, but plays a key role in some forms of death. Perhaps, though, the main reason for not positioning life and death as opposites is because, in many ways, they are not. Additionally, the issue of life-death coevolution (dealt with in part 3) further complicates the life-death relationship. As detailed later, death can enhance life by increasing inclusive fitness in a clonal population and some forms of death played a role in the evolution of more complex life.

While some definitions of life are property based, such that to be alive is to exhibit one or more specific properties, other definitions of life are instead based on processes. Certainly, neither the emergence of life nor of death was a discrete event. In particular, PCD in unicellular organisms is not discrete. It is the progressive extinguishing of the processes uniquely associated with living systems. Not all life processes cease to exist at the same time or at discrete time points. There are, in fact, "degrees of dying" (Durand and Ramsey 2019). From an evolutionary standpoint, the basic fitness components of any organism's life-history strategy (Roff 2002; Stearns 1992), viability and reproduction, may decrease gradually and independently of each other (additional notes 8.1). In unicellular organisms, some of the features of PCD exist even when the organism exhibits others that are associated with life. Spores, for example, are metabolically inactive although this is usually reversible, and senescent cells have lost their reproductive potential, even if they remain metabolically active. In addition to sharing phenotypic features with PCD, the molecular mechanisms of spore formation and senescence overlap with programmed forms of death. In cases where a genetic program for death is implemented, this may lead to PCD, encystation (Khan, Iqbal, and Siddiqui 2015), or cell cycle arrest (Helms et al. 2006; Torgler et al. 1997). It may also be part of a developmental stage (Cornillon et al. 1994), meiotic viral attenuation (Gao et al. 2019), or even the induction of sexual reproduction (Nedelcu, Marcu, and Michod 2004; Nedelcu and Michod 2003). Thomas and others argue that death is a spectrum or range of phenotypes (Thomas et al. 2003). The protein pathways that lead to death are phenotypically plastic. Furthermore, PCD, even in unicellular organisms that harbor such genetic potential, is not inevitable. Not only may the PCD program never be implemented, but in some instances, death is incidental and extrinsic to the cell (for example, predation). As Ameisen highlighted, the term *programmed* is problematic (Ameisen 2002) and it is claimed that the PCD trait is "probabilistic, branching and non-discrete" (Durand and Ramsey 2019).

To understand the meaning of PCD in the unicellular world, a set of terms and definitions is required (Durand and Ramsey 2019) (table 1). The first important distinction is that between programmed forms of cell death and other forms of death, which are due to external factors like physical or chemical damage, lysis by viruses (viral infection can also initiate a PCD-

TABLE 1. Evolutionary definitions of death with examples in unicellular organisms

Concepts of death in unicellular organisms	**Evolutionary definition**	**Examples from the literature**
(True) PCD	PCD is an adaptation to abiotic or biotic environmental stresses resulting in the death of the cell	*E. coli; C. reinhardtii; D. salina; D. discoideum; L. major; B. subtilis*
Ersatz PCD	Ersatz PCD is intrinsic to the cell but has not been directly selected for	*E. coli; D. viridis; D. tertiolecta; P. falciparum*
Incidental death	Incidental death is extrinsic to the cell	Any single-celled organism

Note. PCD, programmed cell death. There are three different "kinds" of death in the unicellular world that are important in evolutionary studies. True PCD is an adaptation to abiotic or biotic environmental stresses, resulting in the death of the cell. Ersatz PCD is also intrinsic to the cell, but the death phenotype itself has not been selected for. Examples include pleiotropy, genetic drift, and trade-offs. Incidental death, for example, physico-chemical damage or predation, is extrinsic to the cell. The same taxon may exhibit PCD or ersatz PCD depending upon the trigger. For example, PCD in *E. coli* is an adaptation to viral invasion and blocks the spread of viruses to others. Ersatz PCD can also occur in *E. coli*, but in this case as a side effect of the mazEF addition module (see the text for discussion). The evolutionary definitions are independent of mechanisms. Autophagy, for example, is adaptive in *D. discoideum* because of the developmental stage of forming stalk structures. However, the same mechanism appears to be pleiotropic in *D. viridis*. Adapted from Durand and Ramsey 2019.

type phenotype but in those instances death is not incidental), or predation. The forms of death that result from extrinsic causes have sometimes been termed *incidental death, necrosis, lytic death, non-PCD,* or sometimes simply *death* (additional notes 8.2). None of these is completely satisfactory to describe all the situations in which programmed forms of death do not occur. Each comes with a history and loaded meaning, but I prefer the term *incidental death* for those cases that have no intrinsic evolutionary history associated with the cell's genetic potential for death.

A second important distinction is the one between PCD and cell aging. This distinction needs to be made explicit because historically the two have sometimes been confused. Before PCD was identified as a distinct entity in the unicellular world, aging and PCD were sometimes considered the same thing. The distinction is not always as clear-cut as we would like, since PCD and aging studies in yeast revealed that the two phenomena can sometimes be intertwined (Fabrizio et al. 2004; Herker et al. 2004). As Herker, Fabrizio, and colleagues have demonstrated in *Saccharomyces,* they overlap in both the genetic mechanisms and phenotypes. In the asymmetric divisions in *S. cerevisiae,* daughter cells bud from a mother cell and after about 20 such divisions the mother cell is sufficiently aged that it loses reproductive potential and viability, and eventually dies, exhibiting some of the features of PCD. But there are fundamental differences between PCD and aging that differentiate them, based on their evolutionary histories. PCD, usually and certainly as it is defined here, does not escape the force of natural selection (Ameisen 2002; Durand and Ramsey 2019) and requires the active transcription of effector genes and the translation of proteins. What will become clear later are the selection pressures that apply to PCD and not to aging. Recently, however, evidence has accumulated to support the hypothesis that aging itself is adaptive in both unicellular and multicellular organisms (Singer 2016). PCD also happens rapidly (hours or at most days) and in most organisms, can be unrelated to how many divisions a cell has undergone. In aging, the process is passive, gradual, pleiotropically linked to other biochemical processes, and sometimes a function of the number of cell divisions. From a general evolutionary perspective, aging and PCD are quite distinct even if the mechanisms do sometimes overlap.

There are two points that are important at this early stage en route to developing a synthesis for the origin and evolution of PCD. The first is that PCD, whatever the genetic program and phenotype (of which there are many), is distinct from incidental death and aging. This is true whether the cell *is* the organism (unicells) or the cell is *part* of a multicellular organism. The second is that PCD has a very different evolutionary and ecological context in multicellular and unicellular life. In multicellular organisms, PCD enhances the fitness of the organism and is essential for normal development and tissue homeostasis (but see chapter 16). In unicellular life, the cell *is* the organism. PCD decreases organismal fitness and, at first glance, does not fit with the "survival of the fittest" aphorism.

How should PCD be defined? The vast majority of researchers define PCD in terms of cellular mechanisms. Defined this way, the evolutionary histories of PCD are not of primary interest. Instead, it is the genetic mechanisms and phenotypic outcomes that determine whether PCD has been realized. Berman-Frank et al. (Berman-Frank et al. 2004), for example, defined PCD as "active, genetically controlled, cellular self-destruction driven by a series of complex biochemical events and specialized cellular machinery" when they studied the demise of species of the marine cyanobacterium *Trichodesmium*. Fabrizio et al. (Fabrizio et al. 2004) define apoptosis (one of the common phenotypes of PCD) as "a form of cellular suicide that leads to the rapid removal of unwanted or damaged cells" and use markers of apoptosis to detect this phenotype in *Saccharomyces*. These mechanistic definitions and measures of PCD were initially introduced from the cell death nomenclature for different phenotypes in multicellular organisms (Kroemer et al. 2005). Unicellular organisms also exhibit "different ways to die" (Jimenez et al. 2009) and the mechanisms and phenotypes (autophagy, apoptosis-like death, paraptosis, ferroptosis, etc.) may vary greatly (see the many examples in Pérez Martín 2008). The natural progression for biologists was to name the kinds of PCD in unicellular organisms after similar cellular phenotypes in multicellular organisms. Understanding these mechanistic differences clearly has an important place, but as Nedelcu and colleagues suggested, the unqualified importing of PCD terminology from multicellular to unicellular life has affected how we understand PCD in the unicellular world (Nedelcu et al. 2011). It has, in fact, resulted in much confusion about what PCD means for single-celled life. Reece et al. also argue that "focusing on the mechanistic differences . . . without the relevant ecological context is not a useful way to progress" (Reece et al. 2011), and Berges and Choi make the point that PCD "interpretation requires clearer definitions of cell death: definitions that are subject to considerable debate even in taxa that are relatively well-explored" (Berges and Choi 2014).

For evolutionary biologists, mechanism-based definitions of PCD are challenging to formulate. There is general "confusion as to how many distinct types of PCD exist" (Reece et al. 2011) in microbes. In some single-celled model organisms, a specific nomenclature for death has been recommended (Carmona-Gutierrez et al. 2018), but even in these cases the mechanistic definitions provide little or no information about the evolutionary history. One of the primary examples of the problem faced is that similar PCD mechanisms and phenotypes may have different evolutionary histories and ecological effects in different taxa. The PCD mechanism, therefore, does not reflect the evolutionary history, and this has led to different interpretations for the same phenomenon. Providing concepts or definitions for PCD based on the evolutionary history is necessary to address this confusion and indicate how the terms are used in this book.

PCD terminology and evolutionary definitions in unicellular organisms

The problems associated with the term *PCD* have prompted some authors to opt for alternative terms that more accurately reflect specific non-incidental forms of death. Ratel et al. (Ratel et al. 2001), for example, suggest *cell death program* and Nedelcu et al. prefer *active cell death* (Nedelcu et al. 2011). The authors' justifications for these terms are not provided here but the arguments they make are valid and the various terms may be even more appropriate. It seems, however, that the more conventional term of PCD is here to stay and introducing new terms at this stage is unlikely to clear up the confusion.

When using the term PCD, I adopt the position developed with Ramsey, which argues for "two main versions of the evolutionary definition" (Durand and Ramsey 2019), one broad and the other narrow. This has become a necessity, since arguments about why PCD evolved often arise because researchers have different concepts of PCD (Berges and Choi 2014). The broad definition of PCD includes any of the mechanisms by which a genetic program for death evolved: natural selection, genetic drift, pleiotropy, and life-history trade-offs (Pepper et al. 2013). The narrow definition is used to investigate conditions where PCD has been selected for. In the narrow definition, PCD is defined as "an adaptation to abiotic or biotic environmental stresses resulting in the death of the cell" (Durand and Ramsey 2019). This we termed *true PCD*. By contrast, *ersatz PCD* refers to instances in which PCD is not an adaptation, but where death still has a seemingly programmed component. Ersatz PCD has not been selected for and is the result of genetic drift, trade-offs, or pleiotropy. From an evolutionary standpoint, therefore, there are two important concepts. First, the narrow definition of PCD, where the trait is an adaptation, and second, ersatz PCD, which includes evolutionary histories that are non-adaptive (table 1).

PCD in multicellular and unicellular organisms

The interest in PCD in multicellular organisms has been primarily mechanism-driven because there is no controversy concerning its evolutionary history. In multicellular life, PCD is explained by kin selection (Gardner, West, and Wild 2011; Michod 1982), which is discussed in more detail in chapter 13. The death of a cell as part of development or homeostasis benefits its clonal relatives by enhancing the whole organism's fitness. It is worth mentioning one instance, however, where this is not the case: cancer. Cancerous cells have gone rogue, such that their evolutionary interests in multicellular organisms diverge from the rest of the organism (Aktipis and Nesse 2013; Heng et al. 2010; Merlo et al. 2006). In this situation, PCD becomes an interesting evolutionary question and mirrors some of the intrigue associated with death in single-celled life. Cancer is an atavism, in that the cell adopts the life history strategy of the ancestral unicellular state. In the same multicellular organism, there are cell lineages with two different evolutionary strategies. The genetic relationship between the cancer clone and the rest of the organism is no longer one of kinship and the two groups of cells compete for nutrients. The non-cancerous cells are all aligned with organismal fitness, but the clone of cancer cells has its own selfish interests at the expense of the organism. However, with respect to the cancer clone itself, PCD can enhance the fitness of the clone of malignant cells (Chen et al. 2014).

The intimate connection between PCD and development in multicellularity was part of the scientific zeitgeist and for some researchers this is still the case. It was counterintuitive that a unicellular life form would actively orchestrate its own demise, since this trait, it was argued, would have been eliminated by natural selection. Early reports of PCD phenotypes in model organisms like *Dictyostelium* (for example, Whittingham and Raper 1960) were frequently contextualized to the evolution of multicellular life (Kaiser 1986; Arnoult et al. 2001). But PCD has subsequently been observed in all the major prokaryote and unicellular eukaryote lineages (see references to individual taxa in Ameisen 2002; Bayles 2014; Bidle 2016; Kaczanowski, Sajid, and Reece 2011; Koonin and Aravind 2002; Lewis 2000; Nedelcu et al. 2011; Pepper et al. 2013; Pérez Martín 2008), and what began as a curiosity had become a "beautiful evolutionary problem" (this phrase was used by an anonymous reviewer of a grant application). Central questions related to this problem include these: What are the mechanisms by which unicellular organisms implement a program for cell death? What are the stimuli for PCD in unicells? And why do they have a genetically encoded mechanism for PCD when it is clearly harmful

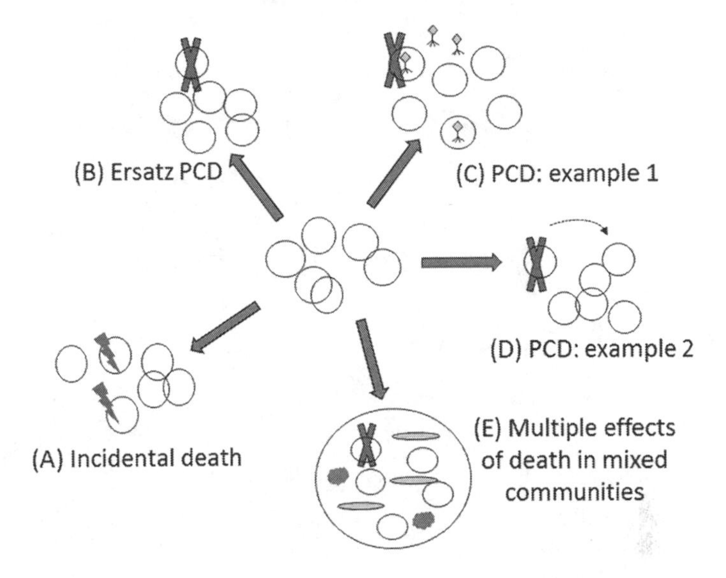

FIG. 17. An evolutionary perspective of the ways in which unicellular organisms die. Mortality in a population of healthy cells may take the following forms. (A) Incidental death, in which cells can be damaged by physical or chemical means and die from extrinsic insults. As a result, the toxic cellular contents are liberated into the external microenvironment and may harm others. (B) Ersatz PCD, in which the cell death phenotype is the result of internal cues, but the mechanism involved is not an adaptation for this death. (C, D) PCD, in which the phenotype is an adaptation for death and evolves by kin/group selection. The mechanisms may vary, and two examples are illustrated. In (C), PCD limits or aborts the spread of viruses through the population. In (D), the fitness advantages are provided by nutritional resources or chemical signals. In (E), microbial communities comprising different taxa may exhibit multiple kinds of death (incidental, ersatz PCD, and PCD) with multiple downstream effects in the community. Adapted from Durand and Ramsey 2019.

to the individual? These issues have resulted in nothing less than a "paradigm shift" in our understanding of mortality in unicells (Ameisen 2002) and the evolution of more complex life (Durand, Sym, and Michod 2016). The evolutionary outcomes of the different ways in which unicellular organisms die have had a profound effect on our understanding of mortality and its role in the living world (fig. 17).

9
Observations of death

One of the issues to arise when PCD was first identified in unicells was whether the methods for investigating the phenomenon were appropriate. This remains a major debating point. The methodological assays used were imported from studies in animals and plants and it is not always clear whether the result means the same thing in unicellular organisms as it does in multicellular ones. From a purely mechanistic point of view, this remains an unresolved question. It is usually recommended that several assays be used in conjunction to confirm PCD and in general, this approach has been suitable, albeit imperfect (see chapter 10). However, the results can still lead to confusion; for example, the use of the TUNEL assay in *Plasmodium* can give false positives (Engelbrecht, Durand, and Coetzer 2012). The unambiguous identification of PCD is also essential for interpreting the evolution-related questions, since the data are useful only if there is certainty that PCD is indeed occurring. Is there a gold standard for demonstrating PCD? This is not only practically important in the laboratory, but the insights gained by examining this question more closely have provided additional clues for developing a general synthesis for the evolution of PCD (chapter 14).

In unicellular eukaryotes (measures of PCD in prokaryotes are covered in chapter 10), direct observation of the cell is perhaps the most conclusive way to determine how it is dying, especially when the aim is to distinguish programmed forms of death (PCD and ersatz PCD) from incidental death. The ultrastructural changes in cellular and subcellular structures reveal whether the way in which a cell is dying includes changes in metabolism, organellar functions, and gene regulation or not. Without a "programmed" component, a cell simply breaks apart in a way that does not include molecular modifications to the cell's architecture. Cells that are killing themselves in non-incidental ways exhibit ultrastructural features that are the result of changes in the molecular mechanisms responsible for cell homeostasis, development and differentiation, and regulatory pathways. The transcription of genes is up- or downregulated and protein pathways are regulated to effect the phenotypic changes observed. At the same time, if viruses are part of the death process, the viral particles can be visualized. Before illustrating the microscopic structural changes that inform our appreciation of PCD, it is worth noting the macroscopic features of cell cultures undergoing PCD. These are themselves revealing and were the initial spark that ignited interest in this field.

Macroscopic features of PCD in cell cultures

Phytoplankton are used as model organisms for many research questions and some of the first macroscopic observations in unicellular PCD research were made in microalgal cell cultures. Depending on the taxon, in liquid culture phytoplankton render the medium a rich green, red, brown, or orange appearance. One of the earliest observations that phytoplankton death was not necessarily a passive process was made by Falkowski, Vardi, Bidle, and others while working with *Emiliania huxleyi* (for an account of the exciting events that led them to their conclusions see Lane 2008). The macroscopic observations were compelling. A culture of *E. huxleyi* seemed to have "dissolved" overnight as opposed to the typical gradual loss of the viability of cells in culture, which is a much longer process resulting in cellular debris and aggregates of non-viable cells. They observed that, overnight, the culture medium had cleared. The chlorophyll that results in the green color of *E. huxleyi* cultures had effectively disappeared (it had been actively degraded by the cells themselves) and all that remained was a sediment at the bottom of the flask. This different way of dying was later discovered to be similar to PCD in multicellular organisms. Until then, PCD was considered a hallmark of multicellularity, and the mechanisms by which PCD occurs in unicellular organisms rapidly became an area of great interest. A similar scenario played out while Berges, Falkowski, Segovia, and colleagues were studying another phytoplankton, *Dunaliella tertiolecta* (Berges and Falkowski 1998; Segovia et al. 2003). The subsequent electron microscopic investigations of these and other taxa revealed that PCD involved the entire remodeling of the cellular architecture and subsequent dissolution of the cell.

The ultrastructure of PCD

The microscopic observations of PCD, in particular the transmission electron microscopy (TEM) studies, laid the foundations for our appreciation of this phenomenon in unicells. Even in the absence of any specific biochemical assays, microscopic observations reveal the complexity of the cellular and subcellular changes indicating that multiple molecular pathways must be playing a part for the features to manifest (fig. 18). Organellar and cellular architectures are remodeled, metabolism is altered, and subcellular components are dissolved, all of which indicate dramatic regulatory changes at the genetic and protein level. Even though there may be different phenotypic manifestations of PCD, the ultrastructural changes reveal unambiguously that the mode of death is not passive cell lysis. This has subsequently been documented in diverse microbes under a range of conditions. For example, in a cyanobacterium living endophytically in a fern (*Azolla microphylla*), different ultrastructural forms of programmed death (apoptosis-like and autophagy-like) were observed, depending upon the stage of the cell's development (Zheng et al. 2013). In *Microcystis aeruginosa*, an apoptosis-like PCD phenotype was induced by exposure to hydrogen peroxide (Ding et al. 2012), and in *Trichodesmium* species, the formation of gas vesicles and vacuoles and the degradation of internal structure (e.g., the thylakoids) were noted (Berman-Frank et al. 2004). The morphological changes that occur in *Dictyostelium discoideum* have been captured with time-lapse video and extensively documented (Levraud et al. 2003). These varieties of death, and the many others that have been documented, must include genetic and protein pathways for the cellular architecture to be modified in this way. There is nothing incidental about these manifestations of death.

Some of the early observations also provided clues as to why PCD may be occurring. *Dictyostelium discoideum* is a social amoeba (social because the organism may live freely or in social groups depending on the environmental conditions) that demonstrates many of the features of PCD that have typically become associated with the phenomenon in metazoa (Arnoult et al. 2001; Cornillon et al. 1994). These include changes like cell shrinkage, membrane contraction and blebbing, vacuolization, and nuclear chromatin condensation. Two other features were observed that have particular relevance for our understanding of the evolution of PCD (discussed further in chapters 11 and 16). The first is the formation of what Arnoult et al. called apoptotic bodies (small membrane-bound vesicles) that were released by the dying amoebae (Arnoult et al. 2001). These structures were observed by both transmission and scanning electron microscopy (SEM). Similar vesicular structures have also been reported in other model organisms dying via PCD, including diverse protists (Deponte 2008) and chlorophytes (Durand, Sym, and Michod 2016). The significance of these vesicles was not clear and has still not been fully un-

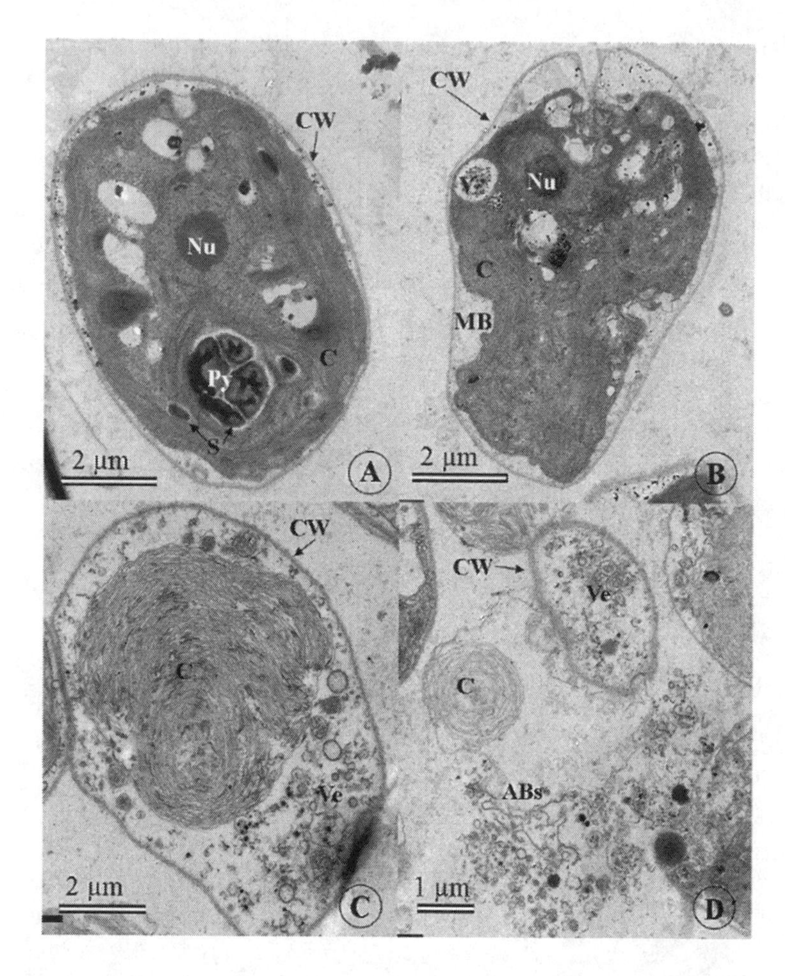

FIG. 18. The ultrastructure of PCD. In incidental death, the cell simply breaks apart and lyses with the intracellular contents liberated into the environment. In PCD, genes are switched on or off and protein pathways are activated that lead to a complete remodeling of the cell and disintegration of its component parts. This example of PCD ultrastructure was obtained following heat-induced PCD. The organism is *C. reinhardtii*. (A) Healthy cell (nucleolus-Nu, cell wall-CW, pyrenoid-Py, chloroplast-C, starch-S). (B) Early (2 hrs) and (C) late (8 hrs) cellular changes after PCD induction. The typical ultrastructural changes that occurred included cell shrinkage, nucleolar condensation (Nu), membrane blebbing (MB), and vacuolization (V). At end-stage PCD, there was near complete dissolution of the chloroplast with stacking of the thylakoid membranes and membrane-bound vesicles (Ve) appeared. (D) After PCD was complete, these vesicles, now resembling the apoptotic bodies (ABs) associated with PCD in multicellular organisms, were noted in the extracellular environment. Image taken from Durand, Sym, and Michod 2016. Different phenotypes of PCD were observed when death was induced by UV irradiation (Moharikar et al. 2006) or nitrogen starvation (Sathe et al. 2019).

covered. They did, however, provide support for those arguing for PCD as an adaptation when it was observed that, in the case of *D. discoideum*, the apoptotic bodies were engulfed by neighboring healthy cells (Arnoult et al. 2001). The suggestion was that natural selection may be occurring at a level other than the single cell, as evidenced by single cells functioning like an integrated group (the "levels of selection" question has been at the heart of the evolution of PCD debate and is covered in chapter 13).

In a completely unrelated taxon, the ultrastructural changes are just as dramatic. *Chlamydomonas reinhardtii* is a unicellular chlorophyte with a cellular architecture very different to that of the amoeba *D. discoideum*. Many of the cellular features associated with plant and metazoan PCD are also noted in *C. reinhardtii*, such as cell shrinkage, membrane blebbing, vacuolization, and chromatin condensation (Durand, Sym, and Michod 2016; Moharikar et al. 2006). In addition, vesicles form from the cellular contents and there can be stacking of the thylakoid membranes and dissolution of the chloroplast, which completely degrades within a day. This is what was also found to occur in the *E. huxleyi* cultures discussed in the previous section. Again, it is not always known exactly why these events occur (although there are several plausible explanations that fit with the experimental data), but the observations themselves provide clues as to the evolution of death.

Why are ultrastructural studies so important in PCD?

The main advantage is that direct microscopy reveals unambiguously whether a cell is dying by PCD or not. This is of special importance for understanding PCD evolution, since it makes the distinction between programmed forms of death and incidental death. For this reason alone, I would argue that direct observation (especially TEM) is the gold standard for confirming whether PCD is occurring, although I expect many researchers may disagree. In addition to the assessment of PCD versus incidental death, there are also many different manifestations of PCD (for example Jimenez et al. 2009; Zheng et al. 2013) and observations of the ultrastructural changes provide additional insights as to the cellular mechanisms. The problem, however, is that TEM is labor intensive, time-consuming, and operator dependent and therefore not always practical in the experimental situation. Light and fluorescent microscopy are easier to undertake and quicker to perform, but not nearly as informative. The non-microscopy methodologies discussed in chapter 10 tend to be much more rapid and often technically simpler, but they are not universally applicable in unicells and should be used judiciously. It should be borne in mind that the biochemical assays for PCD measure only a single feature that is not consistently associated with a particular phenotype. Many researchers have commented how, in the same taxon, a particular PCD phenotype detected by a biochemical assay may be observed in one context but not in another. For example, cells can be resistant or more susceptible to PCD depending on the previous stresses encountered or cellular age (Yordanova et al. 2013). This is possibly epigenetic in nature, but for now the reasons are not known (see postface). As such the biochemical assays may produce false negatives (for example Dingman and Lawrence 2012) or false positives (for example Engelbrecht, Durand, and Coetzer 2012). This has been problematic and even led some researchers to argue that a bona fide PCD molecular pathway has not been adequately demonstrated in some organisms, suggesting that only necrosis or incidental death and viral lysis are the real forms of death (Proto, Coombs, and Mottram 2013). I think this view is easily rebuffed, although it certainly is true that the biochemical assays have limited value and may lead to confusion.

The limitations associated with the biochemical assays and the need for distinguishing PCD from incidental death mean that direct observation of the cell allows for a much more complete interpretation of what is happening. The PCD ultrastructure also reveals features that can be related to specific genes and proteins and in some instances, the molecular machinery has been identified (see chapter 10). For example, the observations concerning membrane biosynthesis, proteolysis, and shifts in metabolic activities point to regulatory activities in lipid metabolism, proteosomal degradation, and carbohydrate metabolism.

From an evolutionary biology perspective, the PCD ultrastructure has provided clues for our understanding of why PCD persists in unicellular organisms when the trait is clearly so harmful. On the one hand, the formation of secretory vesicles (apoptotic bodies), which are engulfed by others in the population, as occurs in *D. discoideum*, provides a mechanistic explanation for the claim that PCD provides group-level advantages. In addition, the switches in gene regulation indicate an active regulated process, suggesting this is not simply a side effect of proteins that are performing unrelated functions. On the other hand, there are some features that may suggest a non-adaptive explanation. The switch to lipid metabolism during nitrogen starvation–induced PCD (Sathe et al. 2019), for example, may simply mean that the metabolic pathways change because of an incapacitation of protein biosynthesis, which requires nitrogen. Following the acceptance by the scientific community that PCD does, in fact, occur in the unicellular world, the primary issue for most biologists was to develop an understanding of why and how this was happening.

10

Mechanisms and measures of programmed cell death in the unicellular world

Our mechanistic understanding of PCD in unicellular organisms, while still rudimentary, has revealed a range of genetic components involved in death. Within a single taxon (especially eukaryote taxa) there is usually more than one molecular pathway leading to death, although there is frequently crosstalk between them. The resultant PCD phenotype (see, for example, the nomenclature for death in yeast in Carmona-Gutierrez et al. 2018) may be morphologically and biochemically distinct or may include features that are associated with more than one morphotype of death. In addition to the variation in the death phenotype itself, it has also become clear that the molecular pathways overlap with other outcomes like cell cycle arrest, dormancy, senescence, aging, spore formation, and sexual reproduction (chapter 8). The complexities of the molecular mechanisms have made dissecting their component parts and attributing functions to them, in a reductionist way at least, very challenging. Nevertheless, significant progress has been made and, in some instances, the molecular basis for a particular death phenotype has been fully dissected. In others, only a few of the key molecules have been identified.

Uncovering the molecular basis for programmed death also provides clues for interpreting the different evolutionary histories and for the levels-of-selection debate. For example, in one taxon, dissecting the mechanism may reveal that it provides no fitness benefits at any level (gene, group, population, etc.), suggesting that the evolutionary history does not involve adaptation, but rather that death is a pleiotropic phenomenon or a side effect of some other function. In another scenario, the molecular mechanism may reveal fitness benefits to clonal relatives, suggesting that kin selection is the explanatory framework in instances where a history of selection is demonstrated. These reasons were the motivation behind the development of evolutionary definitions of PCD and ersatz PCD that are agnostic of the mechanistic and phenotypic vagaries (Durand and Ramsey 2019). The levels-of-selection debate is dealt with later (chapter 13); for now, an understanding of the range of PCD mechanisms and how they are detected is required to support the later synthesis of PCD evolution.

Programmed forms of death in prokaryotes

An impressive diversity of genes and molecular pathways has been implicated in various forms of PCD in bacteria (Lewis 2000) and making sense of what this means in microbial communities has been as much of a challenge as uncovering the mechanisms themselves (Bayles 2014). Despite the general agreements regarding the ways in which bacteria undergo PCD, there are divergent views on how to interpret their evolution. Part of the reason for this, is that until recently there have not been clear concepts or evolutionary definitions for the different kinds of death, like PCD, ersatz PCD, or incidental death (Durand and Ramsey 2019). A good example of this is the toxin-antitoxin (TA) molecular systems, of which there are many types and variants. These are some of the commonest mechanisms of programmed death in bacteria. There are arguments that TAs, like the frequently cited *MazEF* (italicized names refer to genes, non-italicized names to proteins) example discussed later, are the result of population-level selection (reviewed in Bayles 2007, 2014; Engelberg-Kulka et al. 2006). There is also an alternate view that even though TAs are a genetic program and they do lead to cell death, this should not be called or interpreted as PCD at all. The claim is that death is a side effect of selection for the genes themselves (Ramisetty, Natarajan, and Santhosh 2015; Ramisetty and Santhosh 2017). The terminology in chapter 8 (in my view at least) settles the argument by pointing out the different evolutionary histories of the two views. In one ecological context the phenomenon is PCD (it is an adaptation) and in the second it is ersatz PCD (it is a side effect of gene-level selection). Without taking the evolutionary definitions into account and by examining the problem purely from a mechanistic point of view, it is easy to see how conflicting views and interpretations have come about. It is, however, necessary to understand the mechanisms for programmed forms of death in prokaryotes to determine how the program for death is either adaptive or a side effect.

Chromosomally encoded TAs in prokaryotes

MazEF, originally discovered in *Escherichia coli* (Engelberg-Kulka, Hazan, and Amitai 2005; Aizenman, Engelberg-Kulka, and Glaser 1996), has subsequently been found in many bacterial taxa. It is a functional genetic module that is located either on bacterial chromosomes or on extra-chromosomal DNA like plasmids (Moritz and Hergenrother 2007). Chromosomally encoded TAs usually originated at some stage from mobile elements like plasmids but have invaded and settled in the bacterial chromosome. The module codes for a stable toxin (MazF) and a labile antitoxin (MazE). The difference in stability is key to the function. Induction of the module results in the production of both MazF and MazE proteins. Once the module is switched off, however, the toxin outlasts the antitoxin because it is biochemically more stable, resulting in cell death. There are some caveats though: for example, there may be some instances where MazE is overproduced, quenching the long-term effects of MazF, and reversing the path to cell death (Pedersen, Christensen, and Gerdes 2002). But in most instances, death is inevitable. The first important point concerning *MazEF* is that, clearly, it involves a genetic program. This is not incidental death. From an evolutionary point of view, the key question therefore is this: Is this genetic program for death an adaptation at the level of the cell group, as some suggest? In other words, is this true PCD? Or is this ersatz PCD, a side effect of a gene-level adaptation, as others may argue?

TAs encoded on replicons in prokaryotes

MazEF can also be extra-chromosomally encoded on what are, somewhat abstractly, called replicons. Replicons are structurally and functionally distinct from bacterial genomes and may be, for example, plasmids or "minimal" plasmids. From an evolutionary standpoint, the structural and functional separation of TAs, which may be *MazEF* (Moritz and Hergenrother 2007) or many other kinds of TAs (Jensen and Gerdes 1995), from the bacterial genome is significant. When TAs are chromosomally encoded, their evolutionary fate is coupled with that of the genome, whereas the structural and functional separation of the plasmids from the bacterial genome means that the evolutionary interests of the two genetic components are not necessarily aligned. Plasmids are, to greater or lesser degrees, autonomous. Most plasmids are at least "semi-autonomous"—they have their own origin-of-replication even if they may rely on the host's replication machinery. Plasmids are one type of mobile genetic element (MGE), in that they can move horizontally between organisms as well as vertically from parent to offspring. This means that their fate is uncoupled from the bacterium itself.

Replicons such as plasmids, once inside a host bacterial cell, cannot be ejected when they have functional TAs like *MazEF* if the cell is to remain viable. The reason for this is that any cell that loses the plasmid is left with a stable toxin and labile antitoxin, with the stable toxin persisting and leading to cell death. This is also the case when a parent bacterial cell divides by binary fission, producing two offspring. If one of the daughter cells does not have the plasmid, there is no antitoxin produced and the long-lasting toxin leads to cell death. This form of plasmid addiction is called post-segregational killing, and the self-determination of the plasmid's fate means that there is a fitness attachment to the TA module that is distinct from chromosomally encoded TAs, which are entirely under the control of the bacterial genome. In the case of replicon-associated TAs, is this true PCD, is this ersatz PCD, or from the cell's perspective is the plasmid simply an addiction molecule, as sometimes claimed?

Developmental programs and cell death in prokaryotes

Programmed forms of death have also been implicated in developmental pathways in bacteria. An example of the mechanism involved includes the lysis of a mother cell via the secretion of autolysins, such that a spore is liberated. *Bacillus subtilis* bacteria that are defective in the program do not lyse and liberate their spore (Smith and Foster 1995), which in the short term affects the organism's viability and reproductive potential, and in the long term ultimately leads to incidental death. The mechanism responsible for mother cell lysis and sporulation is programmed in nature, since the autolysin is genetically coded, and its activity is regulated by the parent cell. The mother cell's death liberates the spore from within and is part of the organism's developmental lifecycle. Fruiting body formation and sporulation in *Myxococcus* is another example of controlled death that is essential for development. The mechanism involves "autocides," which are enzymes produced after cellular aggregation. The autocides induce most of the cells in cultures of *Myxococcus xanthus* to die, which allows others to form fruiting bodies and sporulate (Rosenberg, Keller, and Dworkin 1977; Wireman and Dworkin 1977). The phenomenon of cell death in this organism is also programmed (Rosenbluh and Rosenberg 1989) and the developmental stages for which they are responsible are essential for reproduction.

In contrast to the TA systems above, there is much less controversy concerning the evolutionary significance of autocide-induced cell death in *M. xanthus*. Since the cultures of mutants that cannot undergo this form of death are much less viable than the wild type, the accepted interpretation is that autolysis is essential for the developmental stage of clonal relatives (or at least very close relatives). Kin selection theory (Gardner, West, and Wild 2011; Michod 1982) is the explanatory framework.

Mechanisms of PCD in unicellular eukaryotes

In contrast to PCD in prokaryotes, the mechanistic basis for programmed forms of death in unicellular eukaryotes is extraordinarily complex. As indicated in chapter 8, this has very often been the result of researchers meaning different things by PCD, but there are also important points to consider at the outset. In many cases, the precise genes and protein pathways are unknown, although in some model organisms the putative pathways have been broadly sketched. In isolated instances, such as the two examples below, there is a much more detailed understanding. As indicated at the beginning of this chapter, the same PCD pathways exhibit a degree of phenotypic plasticity. In addition, pathways in different taxa may be very poorly conserved evolutionarily but still exhibit similar outcomes. The general impression in eukaryotes is that there is considerable crosstalk between PCD pathways and, more significantly for evolutionary interpretations, crosstalk between putative PCD pathways and other vital cellular functions. Those who argue that PCD is predominantly pleiotropic, as opposed to adaptive, have seized upon this aspect. Advocates of PCD as a non-adaptive phenomenon claim that there is very little in the way of true PCD. Rather, what is observed is almost all ersatz PCD. In addition, the difficulty in reconciling the features of PCD with molecular pathways in some groups of organisms like parasitic protozoa have led some researchers to suggest that, at least in these taxa, PCD should not be considered a bona fide phenomenon (Proto, Coombs, and Mottram 2013). Similarly, Voigt et al. (Voigt, Morawski, and Wöstemeyer 2017) "have severe doubts on the existence of an apoptotic program in case [*sic*] of *C. reinhardtii*." These points of view are included here for completeness and to alert the reader that some researchers reject the very idea of PCD, and even ersatz PCD, in unicellular organisms. Clearly, this is not my view. My sense is that, in time, the evidence for PCD (and ersatz PCD) in the unicellular world will become so obvious that even the most ardent critics will relent. There may well be a time in the future when researchers will wonder what the controversy was all about.

True PCD (PCD as an adaptation) and ersatz PCD (non-adaptive PCD), as well as the levels-of-selection question, are each dealt with individually in the next three chapters. From a purely mechanistic standpoint, however, there are some aspects worth highlighting. It is indeed the case that much less is known about the molecular mechanisms in unicellular eukaryotes than in prokaryotes. However, in some model organisms the main components have been worked out and complete pathways are proposed. In addition, the ultrastructural observations discussed earlier confirm the regulated nature and molecular basis for PCD in eukaryotes, including parasitic protozoa. It is possible that programmed forms of death may be absent in some select groups of organisms, but for most researchers there is now no doubt that PCD (true or ersatz) exists in most lineages. The last point worth highlighting before detailing some of the better known PCD mechanisms, is that the molecular mechanisms have been difficult to pin down. The reason is not particularly surprising and alluded to in the first few chapters of the book. Complex traits are difficult to disentangle using the reductionist approach in molecular biology. This is not a problem unique to PCD biology. Organisms are integrated entities and reducing complex phenomena, of which PCD is certainly one, to atomized traits is problematic. Certainly, at the molecular level most traits are polygenic and difficult to dissect. For PCD, there may be key molecules, but the crosstalk between pathways, the variety of environmental factors associated with PCD, the phenotypic plasticity, and the context-dependent realization of the phenomenon all mean that uncovering the molecular basis is always going to be challenging.

PCD in microalgae

The mechanistic studies of PCD in microalgae cover several completely disjunct lineages and include a wide range of environmental triggers. Some examples of the model taxa include species of the genera *Micrasterias* and *Chlamydomonas, Phaeodactylum, Emiliania, Peridinium,* the prokaryote *Trichodesmium,* and many others (for reviews see Bidle 2015, 2016). Many of the central molecular components of the PCD pathways are shared between taxa, which has enabled researchers to develop a putative, generic, mechanistic explanation that captures the key steps.

PCD is induced by a wide range of environmental stresses, including changes in temperature, salinity, light intensity, oxidative stress, nutrient depletion (phosphate, iron, nitrogen, and silicate), as well as viral infection and infochemicals. In natural settings biotic factors can exhibit greater effects on PCD (Kozik et al. 2019). The key steps in the PCD pathway are as follows (fig. 19). The environmental triggers are received by secondary messengers that act as signaling molecules either directly or via the release of calcium stores. These signals lead to the translocation of mitochondrial (cytochrome C) and chloroplast (cytochrome F) molecules (see Zuppini et al. 2009 for an example) from the organelle compartments to the cyto-

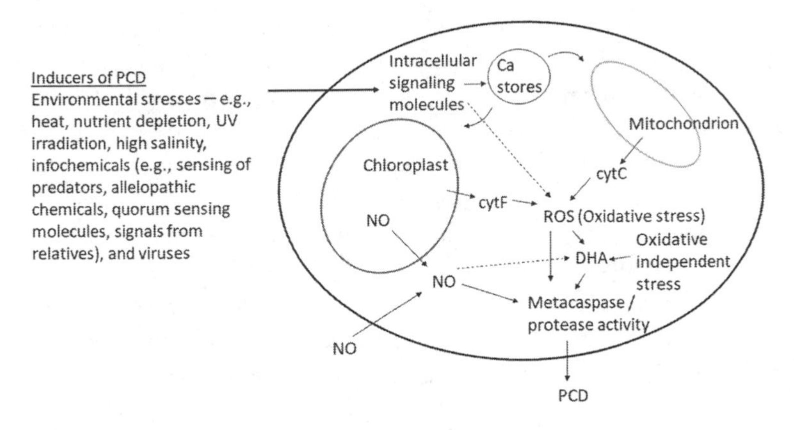

FIG. 19. PCD in unicellular phytoplankton. A broad schematic of PCD in phytoplankton. Only the very basic components and pathways are indicated. See text for details. Redrawn and based on images in Bidle 2015, 2016.

sol, resulting in intracellular bursts of reactive oxygen species (ROS) like H_2O_2 and O_2^-. Nitric oxide (NO), another signaling molecule, can also be produced intracellularly in response to toxic metabolites or diffuse into cells from the environment. The free radicals and molecular messengers interact directly with cellular pathways to upregulate the molecular executioners of PCD-like metacaspases (Mannick et al. 2001). The change in redox potential that results from ROS and NO upregulates antioxidant genes (e.g., catalase, superoxide dismutase, ascorbate peroxidase, dehydroascorbate reductase [DHAR]), which either deal sufficiently with the cellular stress or, depending on their metabolites (e.g., dehydroascorbate, DHA), elicit PCD via metacaspases (Murik, Elboher, and Kaplan 2014). DHA can also activate PCD independent of the ROS pathway, indicating there are mechanisms that lead to PCD via DHA that are unrelated to any burst in oxidative potential (Murik, Elboher, and Kaplan 2014; Vardi et al. 2006). ROS and NO also activate PCD via the post-translational modification of executioner proteins and, in some lineages, lead to the upregulation of death-specific proteins that are responsible for managing the delicate balance between acclimation and PCD (Thamatrakoln et al. 2013). One of the interesting activities of NO and other infochemicals is that they can either induce resistance to environmental stresses in surrounding cells (García-Gómez et al. 2016; Yordanova et al. 2013) or trigger cell death in groups (Vardi et al. 2006). This appears to be dependent upon several variables like the severity and nature of the stress, cellular metabolic activity, and the relative concentration of infochemicals.

The proteins mentioned above are only a select few whose functions have been repeatedly documented. There are many other regulatory proteins and enzymes implicated in both the induction and execution of PCD. Comparative analyses have revealed the evolutionary histories of these so-called domains of death (Aravind, Dixit, and Koonin 1999). Many of them appeared in eukaryotes following the endosymbiosis of the bacterial ancestors of mitochondria (Koonin and Aravind 2002). Homologous forms are present in diverse unicellular eukaryotes (Koonin and Aravind 2002; Nedelcu 2009), which is in keeping with the singular origin of this endosymbiotic event. One of the effectors of PCD that is shared by almost all unicellular eukaryotes is a family of protease enzymes, in particular those in the class C14 peptidases. Phylogenetic and structural studies have resolved the distinct families of PCD-related peptidases, with the ones in microalgae known as metacaspases (Aravind and Koonin 2002; Choi and Berges 2013; Uren et al. 2000) as opposed to the caspases and paracaspases in metazoa and fungi. Notwithstanding the historical confusion and sometimes opposing views concerning the classification of the peptidases, the biochemical mechanisms of action and substrate specificities (Carmona-Gutierrez, Frohlich, et al. 2010; Enoksson and Salvesen 2010) (see additional notes 10.1 and Minina et al. 2020 for the latest on the terminology regarding caspases, metacaspases, and paracaspases), their activities correlate positively with PCD and measuring their activity is frequently used as a proxy for programmed forms of death. There are, however, some discrepancies that have been a source of much controversy. Due to the potential involvement of these enzymes in death in unicellular organisms, a closer examination of their role in phytoplankton is warranted, especially since this controversy has reached to the heart of PCD evolution debate.

Caspases and their homologs

The caspases (acronym for cysteinyl aspartate-specific proteases) are unique intracellular proteases. They are ubiquitous in metazoans (which includes all animals, from sponges to primates), some of which are specifically associated with PCD and other functions in these organisms. The discovery of their mechanism of action in metazoans later became relevant for measures of PCD in unicellular eukaryotes. Caspases have the specific requirement for cleavage after an aspartic acid residue and the recognition of at least four amino acids terminal to the cleavage site. This substrate specificity means that caspase activity in animals is not indiscriminate. It is also true, however, that the caspases not only are active in PCD, they are involved in other cellular processes (Bell and Megeney

2017), and there is also a caspase-independent form of PCD. Detecting caspase activity via their substrates and caspase inhibitors, therefore, is a sensitive marker of PCD (it will detect PCD if caspase-dependent PCD is occurring) but is not specific for PCD (detecting caspase activity does not prove PCD). It was long assumed that detecting caspase activity in microalgae would mean the same thing as it does in metazoa. But this sparked a controversy that has remained, giving succor to the adaptive vs non-adaptive PCD debate.

It transpired that caspases are not encoded in phytoplankton genomes. Instead, there is a family of metacaspases that are phylogenetically distinct proteases found in phytoplankton and plants (Aravind and Koonin 2002; Uren et al. 2000; Jiang, Qin, and Wu 2010). The controversy lies in the finding that the active sites in caspases and metacaspases are markedly different (Choi and Berges 2013; Tsiatsiani et al. 2011; Minina et al. 2017). Metacaspases hydrolyze proteins after arginine or lysine (basic residues), not after aspartate (an acidic residue) as is the case with caspases. The enzymes are both proteases (they fragment proteins), but they do so by very different biochemical mechanisms. With this information, the expectation was that the assays for caspase activity, which make use of substrates specific for caspases and not metacaspases, should never be positive in phytoplankton. But this is not the case. The assays for caspase activity are indeed positive in phytoplankton, a finding that has been reproduced many times and caused much confusion. These two pieces of information (the detection of caspase activity despite their apparent absence in phytoplankton) seem contradictory and have led to robust debate. This is not merely of biochemical or semantic interest; for example, some researchers claim that metacaspases are "definitely caspases" (Carmona-Gutierrez, Frohlich, et al. 2010) while others claim that metacaspases are "definitely not" (Enoksson and Salvesen 2010). It also has implications for interpreting PCD evolution. As indicated above, protease activity is central to PCD, and the debate concerning the assay has gone to the very heart of our understanding of PCD evolution.

A closer examination of the phylogenetic data (Aravind, Dixit, and Koonin 1999; Aravind and Koonin 2002; Choi and Berges 2013; Koonin and Aravind 2002; Uren et al. 2000; Jiang, Qin, and Wu 2010), the biochemical experiments examining the enzyme activities (Sundstrom et al. 2009; Choi and Berges 2013; Tsiatsiani et al. 2011), and the numerous findings of phytoplankton demonstrating caspase-like or metacaspase activity (for example, Bar-Zeev et al. 2013; Berges and Falkowski 1998; Bidle et al. 2007; Orellana et al. 2013; Segovia et al. 2003; Wang et al. 2017) lead to conclusions that assist with the future interpretations concerning PCD evolution. It is clear that the caspase assays are detecting cysteinyl aspartate-specific protease activity in phytoplankton and the results are not false positives. This has been shown in multiple taxa, the substrates are specific, and the data are reproducible. How is this possible in taxa that do not have caspases, but instead encode metacaspases? Those arguing against PCD as a bona fide trait in phytoplankton have seized on this discrepancy, suggesting that the interpretations of what PCD is in unicellular organisms are flawed because the assays to detect it cannot be trusted. The rebuke, however, is that there are other proteases in phytoplankton that have cross-reactivity with the caspase substrates. In other words, the caspase-like activity being documented in algae is due to proteases that cleave the caspase-specific substrates but are themselves not homologous to the caspase family. Such enzymes have been found in plants (Vartapetian et al. 2011), and I suspect the same will emerge in phytoplankton, in particular the green algae. It is also possible that there is a second domain in metacaspases, distinct from the active site that has caspase-like activity (Bidle and Falkowski 2004; Uren et al. 2000). It is also worth noting that metacaspase activity was significantly and positively correlated with caspase-like activity in the cyanobacterium *Trichodesmium* (Spungin et al. 2019). The peptidase controversy aside, there is no doubt that metacaspases are (usually) upregulated and play a central role in PCD in some phytoplankton (Bidle and Bender 2008; Bidle et al. 2007) and yeast (Watanabe and Lam 2005; Silva et al. 2005). The metacaspase-specific substrates have demonstrated this (Minina et al. 2017) and the presence of metacaspases in microalgae has been confirmed with genomic studies (for example, Armbrust et al. 2004; Merchant et al. 2007).

PCD in yeast

Saccharomyces cerevisiae (baker's yeast) is a model unicellular eukaryote used in a wide range of research programs and some of the earliest PCD mechanistic studies made use of this organism. Some of the PCD machinery identified in phytoplankton is also found in yeast. The signaling molecules ROS and NO (as well as ammonia, which is not the case in phytoplankton) also play a central role. ROS were identified very early on as being a key mediator in several PCD pathways (Madeo et al. 1999) and their concentration may increase in response to external environmental stress or internal cellular signals. NO is also an inducer of death. It can be produced endogenously and decreasing the NO concentration promotes cell viability. The role of ammonia, a metabolic by-product in aging yeast cells, is particularly interesting because of its group-level effects (this is discussed in detail in the next chapter). The accumulation of ammonia signals to healthy cells to switch metabolic activities, which minimizes oxidative stress and allows them to survive. Older, less healthy cells are less capable of minimizing the oxidative stress and undergo PCD (Vachova and Palkova 2005).

The yeast model system has been more amenable to molecular dissection of the PCD machinery and has revealed a complex interplay between molecules involved in signaling pathways. Proteins involved in PCD are finely regulated with activators, inhibitors, and executioners interacting with each other in response to external and internal (both genetic and epigenetic) triggers. The mitochondrion is intimately associated with PCD in yeast (Carmona-Gutierrez, Eisenberg, et al. 2010; Ludovico, Madeo, and Silva 2005), although there are also pathways independent of this organelle. Disruption of mitochondrial activity or the activation of regulatory proteins can lead to the translocation of cytochrome C (cyt C) from mitochondria to the cytosol (for example, Manon, Chaudhuri, and Guerin 1997) and an increase in ROS. As in phytoplankton, one of the executioners (there are others) that is activated in response to ROS and cyt C, is yeast metacaspase (Madeo et al. 2002), which is responsible for many of the markers used to detect PCD.

Measures of PCD in unicellular eukaryotes

The argument was made earlier (chapter 9) that the gold standard for determining whether a programmed form of death is occurring is direct microscopic observation. Electron microscopy can also be used to quantify the proportion of cells in a population dying by PCD (for example, Moharikar et al. 2006). However, this approach is seldom practical. In most experimental designs, it is not feasible to wait for confirmation of cell death by TEM before proceeding to the next step. How then can one be sure when PCD (either true or ersatz) is occurring? An understanding of how PCD is measured is necessary for interpreting results and deciding when PCD is adaptive and when it is not, and for developing a general synthesis of PCD evolution.

In time, it is likely that definitive (either causal or very highly correlated) markers of PCD in different taxa and in response to different stimuli will emerge. Until then, several markers are used to give an overall assessment of what is happening. It is true that there are sensitive markers of PCD that are causally related to the implementation of the death program, but these can be troublesome. As many researchers have noted, the markers of PCD were imported from the literature regarding multicellular organisms. However, the evolutionary implications of PCD in multicellular and unicellular organisms are very different and positive assays for PCD in the unicellular world may not mean the same thing (for example, Arnoult et al. 2001; Berges and Choi 2014; Bidle 2016; Carmona-Gutierrez, Eisenberg, et al. 2010; Debrabant and Nakhasi 2003; Durand and Ramsey 2019; Durand, Sym, and Michod 2016; Franklin, Brussaard, and Berges 2006; Kaczanowski, Sajid, and Reece 2011; Klim et al. 2018; Nedelcu et al. 2011; Ramisetty, Natarajan, and Santhosh 2015; Reece et al. 2011; Bayles 2014). Furthermore, even when the markers are used, it is not always clear how reliably they reflect PCD.

To illustrate the limitations of the measures of PCD in the unicellular world, consider the following:

(i) An assay that is 100% sensitive (if PCD is occurring it will be detected, but it may also be positive in other cases) but not 100% specific (the assay may miss cases where PCD does occur) does not prove PCD. The metacaspase / caspase assays typically fall into this category (Seth-Pasricha, Bidle, and Bidle 2013; Teresa Mata et al. 2019) as does the assay for oxidative stress. For example, detecting an increase in ROS is a sensitive marker of PCD. If it is positive it may be detecting PCD. But it may also be positive because of other cellular processes like acclimation or reproduction that are unrelated to PCD (Nedelcu, Marcu, and Michod 2004; Segovia et al. 2015).

(ii) ROS detection and its increase or decrease mean different things in different ecological contexts. ROS may be pro-death or pro-life (Foyer 2018; Zuppini, Gerotto, and Baldan 2010).

(iii) An assay may be neither very sensitive nor specific, which can mistake incidental death for PCD. This has been reported to occur with the TUNEL (terminal deoxynucleotidyl transferase dUTP nick-end labeling) assay (for example, Engelbrecht, Durand, and Coetzer 2012), which detects single- and double-stranded DNA nicking. The issue is that the assay detects DNA damage that may or may not lead to PCD. DNA damage can be a trigger for PCD and not necessarily the result of PCD, although the assay is also positive once there is DNA laddering in the later stages of some kinds of PCD. The same limitation would occur in other methodologies for detecting DNA damage (Kumari et al. 2008). There may be PCD mechanisms that are independent of caspase-like or metacaspase activity in yeast (Madeo et al. 2009). In other words, the caspase or metacaspases assays may not be positive in all cases of PCD.

(iv) The assay sensitivity and specificity change depending on a particular context, which has led to discussions concerning assay reproducibility. One kind of environmental PCD trigger, for example, causes mitochondrial membrane depolarization (a marker of PCD) in *Chlamydomonas*, while another does not (Vavilala et al. 2015). The environmental triggers for PCD also depend on the age or growth stage of the culture and previous exposure to stress can render cells more or less susceptible to PCD (Affenzeller et al. 2009; Segovia et al. 2003; Segovia et al. 2015; Yordanova et al. 2015).

(v) Intercellular communication and environmental signaling between individuals in a population affect a cell's susceptibility to triggers of PCD (Vardi et al. 2007; García-Gómez et al. 2016; Yordanova et al. 2013).

(vi) Markers of PCD differ between taxa even when the same stimulus is used.

(vii) The caspase assays (see the discussions above and chapter 12) are controversial in phytoplankton. Caspase-like activity is regularly detected but the caspase enzymes themselves are absent. Only metacaspases, which have very different substrate specificities, have been identified in phytoplankton.

(viii) The annexin V assay, which is used in mammalian cells to detect phosphatidylserine (PS) exposure on the outer leaflet of the cell membrane (a marker of the loss of membrane asymmetry that occurs in PCD), is used extensively in unicellular organisms. The assay appears to work well in most taxa, but it was not expected to produce positive results in model organisms that lack PS like *Leishmania* (Weingartner et al. 2012) and *Chlamydomonas* (Giroud and Eichenberger 1988). Despite this, apoptotic promastigotes in *Leishmania* do bind annexin V (van Zandbergen et al. 2006), and chlorophytes like *Dunaliella* (Orellana et al. 2013) and *Chlamydomonas* (Durand et al. 2014; Moharikar et al. 2006; Voigt and Woestemeyer 2015) that die by PCD have been shown repeatedly to bind annexin V although it is not always clear if this is on the inner or outer leaflet of the membrane. It seems that annexin V is not as specific for PS as originally thought, but also binds phosphatidylethanolamine and phosphatidylglycerol (Weingartner et al. 2012), which are present in the two taxa above. This would explain the reported findings, but not knowing the mechanistic basis for this assay fuels the uncertainty. In addition, it has been suggested that heat-induction of PCD may disrupt the membrane in some organisms, resulting in the exposure of phospholipids in the absence of PCD (Dingman and Lawrence 2012). In these instances, a corroborating assay is required.

The confusion concerning the interpretation of the measures of PCD in unicellular organisms has prompted some authors to suggest that particular terminologies be used for specific organisms (Carmona-Gutierrez et al. 2018), that evolutionary concepts be employed instead of methodological ones (Durand and Ramsey 2019) or that the interpretations be contextualized to ecological conditions (Berges and Choi 2014; Reece et al. 2011). This is a welcome advance, but in most instances, there is still insufficient evidence to interpret the meaning of the death phenotype and classify it according to the methodological, taxonomic, or evolutionary definitions. Routine markers will continue to be used and considering the very real problems associated with detecting PCD in the unicellular world, how can the results be interpreted when trying to develop a more general understanding of PCD evolution? This is a question often raised by those who are critical of cases where PCD is sometimes interpreted as an adaptation. The argument is that PCD is not demonstrated conclusively and it is unknown how the cell is dying.

Clearly, there are significant problems in detecting and interpreting PCD in some unicellular organisms. Any responsible scientist would agree that where there is uncertainty as to what is being observed, these data should not be used to develop a general understanding of PCD evolution. At the same time, however, it is easy to overstate the problems, and extrapolating the obstacles and limitations in some methodologies to the entire field of unicellular PCD is unreasonable. Even with the methodological issues, in many cases there is no confusion about what is being observed. The ultrastructural studies discussed earlier reflect unambiguously when death is not incidental and occurring via PCD, even if the mechanisms are not clear. In other cases, the molecular mechanisms have been characterized in excruciating detail and PCD is studied with adequate controls and without any other influences that may impact the empirical findings. Which data then are useful when deliberating on the adaptive versus non-adaptive PCD debate and for generating a synthesis of PCD evolution?

The gold standards, hard and soft signs of PCD

As the field has progressed, I have used a general classification for interpreting the markers of PCD (table 2). A gold standard for demonstrating any cellular or molecular trait is a method(s) that illustrates unambigu-

TABLE 2. Measures and markers of PCD

Markers of PCD	Interpretations	Significance
Transmission electron microscopy	Changes in ultrastructure are typically characteristic of PCD	Gold standard
Definitive molecular characterizations (e.g., TA modules)	The genetic basis for PCD is established in some model systems	Gold standard
DNA laddering by gel electrophoresis	DNA laddering is the result of endonuclease activity, which is very specific for PCD	Hard sign
Ejection of the nucleus	Ejection of the nucleus is found only in programmed forms of death	Hard sign
Loss of membrane asymmetry	Loss of membrane asymmetry is specific for PCD but there remain questions concerning the assay used (annexin V)	Hard / soft sign?
DNA (double or single strand) nicking	This form of DNA damage is non-specific and found in PCD and other conditions	Soft sign
Upregulation of PCD-associated genes	Many PCD-related genes are not specific for PCD and are associated with other functions	Soft sign
Caspase, caspase-like, or metacaspase activity	These enzymes are required for most kinds of PCD, but are not specific to PCD	Soft sign
Light microscopy	The cellular changes associated with PCD are not always visualized by light microscopy	Soft sign
Mitochondrial depolarization	This marker is typically positive during PCD, but it is not clear how specific it is	Soft sign
Increase in reactive oxygen species (ROS)	ROS plays a role in most PCD mechanisms, but they are non-specific and associated with other stress responses (e.g., acclimation)	Soft sign

Note. PCD, programmed cell death.

ously that the trait is present and that there is no other reasonable explanation for the finding. It is the measure against which other assays are compared and is simultaneously the most sensitive (no false negatives) and most specific (no false positives; if the assay is positive, it excludes other conditions) (additional notes 10.2). For PCD in unicellular eukaryotes, this takes the form of direct observation (transmission electron microscopy, TEM) as discussed in the previous chapter. Of course, observation is operator dependent and there is always the concern of artifact and the data still require interpretation, but for experienced microscopists this is usually not problematic. The unambiguous molecular measures of PCD like the TA modules discussed above or the Abi genes (chapter 10) in prokaryotes are also gold standards. With time, molecular markers in eukaryotic microbes will be uncovered and potentially become the gold standard but, judging by the reports of PCD in the literature, it appears that most researchers would agree on the importance of direct observation in eukaryotes. The first demonstrations of PCD in any taxon typically rely on ultrastructural observations, against which other "hard" and "soft" signs (see below) are compared. Once PCD is proven via TEM, direct observation is sometimes not repeated because it is a labor intensive and time-consuming process, and workers make use of established triggers of PCD and the less laborious biochemical assays. Alternatively, in cases where the molecular process has been explicitly demonstrated and is not associated with any function unrelated to PCD, one can make use of a marker of the molecular process to conclude that PCD has been definitively demonstrated. In prokaryotes, this is often the case because the molecular mechanisms have been thoroughly worked out and PCD cannot be confused with anything else (Engelberg-Kulka, Hazan, and Amitai 2005; Refardt, Bergmiller, and Kümmerli 2013). In these cases, observation is unnecessary and a molecular assay is used as the gold standard.

The "hard signs" category is reserved for markers of PCD that are highly specific for PCD, but not especially sensitive. If the hard sign is positive, it is almost certain that PCD is occurring. If the assay is negative, it does not exclude PCD. There are two good examples of this. The "ejection of the nucleus" phenotype indicates that cell death is not incidental. The cell actively ejects the nucleus while undergoing PCD (Orellana et al. 2013), but this trait seldom occurs. So, while it may be a specific marker of PCD, it is not sensitive in picking up all the possible cases of PCD. Similarly, double-stranded DNA digestion that generates fragments of particular lengths (the "DNA laddering" phenotype) indicates endonuclease activity. This marker indicates the active enzymatic cleavage at sites between nucleosomes, which does not occur during incidental death. Detecting this phenotype, however, is technically difficult because the process does not seem to be associated with a definitive timeframe or PCD trigger. DNA laddering is also not (usually) found in dinoflagellates because of the genomic architecture. Nevertheless, when DNA laddering does occur it is a strong signal of PCD. But when the marker is absent it does not exclude PCD.

The "soft signs" of PCD include a range of markers and assays that are sensitive for detecting PCD when it is occurring, but that are not specific. The upregulation of genes like superoxide dismutase, catalase, and ascorbate peroxidase that occur in PCD is an example. But these genes are not specific for PCD and may be upregulated in other conditions. The assays for metacaspase activity, ROS, or mitochondrial depolarization also fall into this category. Markers of DNA damage, like the TUNEL assay, which detects single- or double-stranded breaks, are positive whenever there is DNA damage leading to nicking in the genome; this is a sensitive marker but not unique to PCD.

The detection of phosphatidylserine (PS) externalization in the cell membrane appears to be a specific marker of PCD. However, there are controversies concerning its usage because it is also positive in some organisms that do not have PS in the membrane (*Leishmania*, *Chlamydomonas*, and possibly many others). It appears that this may be due to the detection of other phospholipids (phosphatidylethanolamine or phosphatidylglycerol) as well as PS by annexin V. The loss of membrane asymmetry is the important marker (only healthy membranes exhibit this asymmetry), so it is not crucial which phospholipid is being detected to identify PCD. However, there is still some confusion as to what a positive annexin V assay means. Furthermore, false positives are possible. In experiments with *Ankistrodesmus* (a unicellular chlorophyte) the assay can be positive in groups of cells in the absence of PCD. This is presumably due to the annexin V becoming trapped in the extracellular matrix holding the cells together (personal communication, Baretto Filho).

Which markers of PCD should be used and how should they be interpreted?

The utilization of PCD markers depends on the aim of the study. If a simple documentation of the occurrence of a programmed form of death is the aim, then TEM and the ultrastructural changes as well as the molecular genetic markers are most reliable. In conjunction, it may be helpful to correlate some of the biochemical markers with the TEM and genetic markers. It can then be determined whether the biochemical markers are indeed positive, and under what circumstances. They can then be used in experimental set-ups where TEM is impractical. In instances where PCD has been documented and repeatedly demonstrated in association with biochemical markers, then simply using the biochemical markers that are known to be accurate under a particular set of experimental conditions is reasonable. It should be borne in mind, however, that PCD is a phenotypically plastic trait and investigating PCD without a biological context is problematic. Of the biochemical markers, some are allocated more importance than others. The hard signs, when positive, are a very strong indication that PCD is occurring. The soft signs are not specific enough to be used on their own, even if several are positive at the same time. It is my opinion that, in the absence of a gold standard of PCD (TEM or where the molecular characterization is known, for example as in Engelberg-Kulka, Hazan, and Amitai 2005; Refardt, Bergmiller, and Kümmerli 2013), at least one hard sign is required together with one or, preferably, two of the soft signs to conclude that PCD occurred. This is the guide I have used when analyzing and interpreting the data concerning PCD evolution.

11
True PCD
WHEN PCD IS AN ADAPTATION

The spark that ignited the interest in PCD evolution in unicellular organisms concerned the question of why PCD may have evolved by natural selection. What, if any, fitness advantage can be conferred by a trait that is clearly so deleterious? To answer this question, one needs to consider all the possible levels at which selection may act, including gene, cell, group, kin, population, and species-level selection (chapter 13 is dedicated to the levels-of-selection question and the meanings of kin, group, or population; see also additional notes 11.1). Most of the experimental data that support the argument that PCD is an adaptation in unicellular organisms focus on kin-, group-, or population-level selection. As Zuppini et al. say, the explanations for PCD being selected for, are "based on the concept that unicellular life could be able to organize itself into co-operating groups" (Zuppini, Andreoli, and Baldan 2007). These are the data that need to be highlighted to examine the evidence that PCD is an adaptation (the issue of what is an adaptation is returned to later in this chapter)—in other words, that it is an altruistic trait. A series of pointed questions can be asked to interrogate the experimental data and determine whether PCD is, in at least some circumstances, an adaptation (Durand and Ramsey 2019).

What are the proposed mechanisms by which PCD may be adaptive?

There have been at least five ways proposed by which PCD may be adaptive, and there are varying degrees of empirical support for each. First, in parasites PCD has been considered a mechanism for controlling parasite density in the host, thereby increasing host survival and favoring parasite transmission (Al-Olayan, Williams, and Hurd 2002; Debrabant and Nakhasi 2003; Deponte 2008; Engelbrecht and Coetzer 2013; van Zandbergen et al. 2010). Second, in populations of unicellular organisms, it was proposed that PCD limits the spread of infection by viruses, which increases population viability (Hazan and Engelberg-Kulka 2004; Vardi et al. 2012; Vardi et al. 2009). Third, PCD has been documented playing a critical developmental role for group and sometimes multicellular-like behavior (Bayles 2007; Cornillon et al. 1994; Engelberg-Kulka et al. 2006). Fourth, PCD can be a way of sharing resources during times of nutrient depletion (Bar-Zeev et al. 2013; Franklin, Brussaard, and Berges 2006; Arnoult et al. 2001), and fifth, in response to physiological stress (nutrient depletion as well as other environmental stressors) populations can regulate their own growth by release of infochemicals (Yordanova et al. 2013; Zuo et al. 2012) or quorum-sensing molecules (Kolodkin-Gal et al. 2007). The proposed mechanisms that indicate a benefit to others or that regulate cell density, however, are on their own insufficient to determine whether PCD is selected for. It is important to provide evolutionary explanations that can be tested, without which the arguments remain at the level of "naive group selection" thinking (Williams 1966).

What are the proposed evolutionary explanations for PCD?

As indicated above, the evolutionary explanations for PCD being selected for are rooted in our understanding that unicellular organisms are sometimes social beings and may live in cooperative groups. The question of whether PCD is selected for is examined later. Here the positive effects of PCD are discussed.

Some of the earliest indications that PCD positively impacts others in the group came from the model unicellular eukaryote *Saccharomyces cerevisiae* (Fabrizio et al. 2004; Herker et al. 2004) and the prokaryote *Escherichia coli* (Hazan and Engelberg-Kulka 2004). Herker and others found that in *S. cerevisiae* "old yeast cultures release substances into the medium that stimulate survival of other old cells" (Herker et al. 2004) and Fabrizio et al. discovered that the substances promote "the regrowth of a subpopulation of better-adapted mutants rather than life span extension in the surviving population" (Fabrizio et al. 2004). These data led to a possible evolutionary explanation for PCD by suggesting that there could be selection of cooperating groups that include individuals dying by PCD. At this stage, however, the data and explanation did not separate aging from PCD, and the level of selection was not made clear or explicitly tested.

Hazan and Engelberg-Kulka invoked the "characteristics of multicellular organisms" in bacterial cultures to demonstrate that the costs of death at the individual cell level can be offset by selection between populations (Hazan and Engelberg-Kulka 2004). Kolodkin-Gal and colleagues suggested that *MazEF*-mediated death could be a population-level phenomenon, where the mechanism involves a quorum-sensing molecule (Kolodkin-Gal et al. 2007). However, while the potential mechanism was clearly demonstrated, selection was not. The levels-of-selection issue was again not explicit, although the argument was that the toxin-antitoxin (TA) mechanism for death in these experiments was a form of PCD, which benefited others in the population (Hazan and Engelberg-Kulka 2004; Hazan, Sat, and Engelberg-Kulka 2004; Kolodkin-Gal et al. 2007). An additional point of contention is that the TA mechanism is not universally accepted as a form of PCD (see chapter 10). This is the view of Ramisetty and others (Ramisetty, Natarajan, and Santhosh 2015; Ramisetty and Santhosh 2017) and Ameisen interprets TA mechanisms as addiction molecules without the need to invoke higher levels of selection (Ameisen 2002). I agree with a variant of Ameisen's interpretation and will discuss this in more detail in chapter 13.

Has a direct fitness comparison on others in the population been performed between PCD, incidental death, and no death?

The fitness effects of PCD on others in the population have been compared to those due to cellular lysate or no death and it was found that "how an organism dies affects the fitness of its neighbors" (Durand, Rashidi, and Michod 2011). Cell lysate was used as a proxy for incidental forms of death like necrosis, where cells are damaged physically or chemically and the contents leak into the microenvironment. In this study, others in the population produced more offspring when exposed to the supernatant of cells dying by PCD compared to the supernatant of healthy cells. Cell lysate was harmful. Similar benefits were demonstrated in *Dunaliella salina* (Orellana et al. 2013) and again in *C. reinhardtii* cells following induction of PCD by the toxic anti-metabolite mastoparan (Yordanova et al. 2013). In the latter case, the benefit was not that those exposed to the PCD supernatant produced more offspring but that they became more resistant to death. Population-level fitness differences are also associated with PCD in yeast colonies and in *Leishmania major*. In yeast, ammonia accumulates in the center of the colonies and triggers the death of older cells, "allowing young cells on the rim to exploit the released nutrients" (Vachova et al. 2004; Vachova and Palkova 2005). Knocking out the transcription factor Sok2p results in the inability of cells to produce ammonia. This leads to diffuse death throughout the whole population and diminishes the life span of the colony. Colonies that produce ammonia have increased viability. A similar increase in population viability was demonstrated in *L. major*. The entire population lost viability if it was depleted of cells dying by apoptosis-like PCD, indicating that "apoptotic promastigotes, in an altruistic way, enable the intracellular survival of the viable parasites" (van Zandbergen et al. 2006). These data showed that "controlled and selective cell death confers fitness advantages in unicellular organisms" (King and Gottlieb 2009). PCD can be altruistic, but what remained was to determine whether there is selection at this higher level.

Can PCD be explained by kin or group selection?

The individuals used in the experimental populations of *Saccharomyces* (Herker et al. 2004), *Chlamydomonas* (Durand, Rashidi, and Michod 2011; Yordanova et al. 2013), and *Dunaliella* (Orellana et al. 2013) were clonal, or at least very close genetic relatives, allowing for spontaneous mutations in culture. In addition, in *C. reinhardtii* PCD is negatively allelopathic and harms other species (Durand et al. 2014). In the instances of clones or close genetic relatives, the theory of kin selection (Michod 1982; Maynard Smith 1964) easily explains how costly individual behaviors can evolve if the cost/benefit ratio is less than the degree of relatedness (Hamilton 1964). This is certainly feasible and PCD has been shown, theoretically at least, to evolve by kin selection alone (Vostinar, Goldsby, and Ofria 2019). Kin and group (as well as some multilevel) selection approaches can be treated as functionally equivalent (Lehmann et al. 2007), although they are not causally the same. The explanatory framework for PCD as an altruistic trait, therefore, is that it evolved by kin selection in clonal or very closely related populations or by group selection where relatedness is low. Sometimes, however, it is not clear what the "group" is. In *S. cerevisiae*, a mutant subpopulation benefited preferentially from PCD (Fabrizio et al. 2004) and the increase in group fitness can occur even when the group does not comprise clonal relatives (Refardt, Bergmiller, and Kümmerli 2013). In the latter case, PCD was maintained by natural selection in response to viral infection, provided that the starting frequency of PCD in the population was high. A mathematical model also predicted the evolution of PCD (and aggregation) as an optimal strategy in response to viral invasion of the population (Iranzo et al. 2014).

In the *L. major* experiments, populations with PCD survived while those without PCD lost viability (van Zandbergen et al. 2006; van Zandbergen et al. 2010). The genetic relatedness between individuals was not measured in this instance and the mathematics not worked out, so it is not known whether this could be explained by kin or group selection.

Are there any in vivo or field data on PCD?

For eukaryote organisms especially, in vivo or field data would provide further support, since "laboratory microorganisms that have been cultured for long periods under optimized conditions might differ markedly from those that exist in natural ecosystems" (Palkova 2004). Many of the experimental model systems, like *Chlamydomonas* and *Saccharomyces*, have been in culture for many years; however, some experiments were conducted on freshly sampled, naturally occurring organisms. The phytoplankton-archaeon system (Orellana et al. 2013) was isolated from the Great Salt Lake, USA, and the experimental results in yeast cells were confirmed from organically grown Californian red grapes (Fabrizio et al. 2004). In addition, the dinoflagellate *Peridinium gatunense* used to study PCD synchronization in populations was isolated from Lake Kinneret, Israel (Vardi et al. 2007). Although these data were obtained from freshly isolated organisms from the field, the experiments were carried out under controlled laboratory conditions. Data for direct measurements in the field are scanty, but recent work by Spungin, Bidle, and Berman-Frank (2019) documented metacaspase activity in *Trichodesmium* in response to iron depletion (a well-known trigger of PCD) in the south Pacific Ocean. There are also very little in vivo data, with the notable exception of the *Leishmania major* studies, which were carried out in animals as well as in vitro (van Zandbergen et al. 2006).

Although the field and in vivo data are isolated cases, it does seem reasonable to assume that at least some of the results from the fitness and selection experiments above can be extrapolated to natural settings.

What can be concluded?

In many instances, PCD has a positive effect on fitness in others and in some cases the precise mechanism has been observed, for example, the release of apoptotic bodies in dying amoebae and their uptake by healthy ones (Arnoult et al. 2001). The data have also demonstrated that in some cases kin / group selection explains the maintenance of the PCD trait, but an understanding of what is meant by adaptation is also required to appreciate the definition of true PCD.

The meaning of "adaptation"

There have been many debates over how to understand adaptation (Gardner 2017; Gould and Lewontin 1979; Gould and Lloyd 1999; Gould and

Vrba 1982; Hendry and Gonzalez 2008; Reeve and Sherman 1993; Rose and Lauder 1996; Sansom 2003; van Valen 2009; Williams 1966). A distinction can be made between traits that are adaptive (these have a current fitness benefit) and those that are an adaptation (those that are due to an evolutionary response to past selection for the trait) (van Valen 2009). In the first instance, some biologists are satisfied with the ahistorical conception of adaptation, in which an increase in fitness causally related to a trait is sufficient to infer that the character is an adaptation (Reeve and Sherman 1993). In other words, if the trait is adaptive, this is sufficient to conclude it is also an adaptation. For PCD, this interpretation is easily satisfied as many of the experiments above illustrate. The more widely accepted view, however, is that a trait is an adaptation only if it has a particular evolutionary history. This evolutionary history, as Williams argues, must involve the trait exhibiting a demonstrable fit to some function and that the function confers a selective advantage (Williams 1966). In this second instance, the PCD experiments require more detailed consideration and rely on demonstrating that fitness effects of PCD on others in the population are selected for. Obtaining experimental evidence for this interpretation is often very difficult in the laboratory setting, but some of the available PCD data do help resolve this issue.

Perhaps the strongest evidence comes from the phage-induced PCD experiments in *E. coli* (Refardt, Bergmiller, and Kümmerli 2013). These experiments demonstrated that PCD confers a group-level fitness advantage, that the molecular basis for PCD is linked to PCD only and to no other trait, and that a group with the PCD trait outcompetes one without. The group with altruistic death was selected for. For most, if not all, evolutionists, this fulfills the criteria for PCD as an adaptation. However, some researchers may argue that this highly regulated experimental design may not be appropriate for unicellular eukaryotes (as opposed to prokaryotes like *E. coli*), that the experiments were only demonstrated in vitro, or that there should also be a fitness comparison between PCD and incidental death.

The PCD experiments conducted in the eukaryote *L. major* (van Zandbergen et al. 2006) were performed in vivo using laboratory mice. The potential of the organism to cause disease was used as a proxy for fitness, which seems reasonable, since virulence is measured by viability and reproduction. In this instance the populations were isolated in different hosts (strong population structures) and the argument was made that populations with the PCD trait were selected for because the populations without PCD demonstrated a decrease in viability (van Zandbergen et al. 2010).

The work in bacterial biofilms (Bayles 2007; Engelberg-Kulka et al. 2006) did not include selection experiments; however, PCD was important for biofilm development (Rice and Bayles 2006; Bayles 2007) and multicellular behavior (Engelberg-Kulka et al. 2006). The multicellular stage and development of spores in *Dictyostelium* was also contingent on the PCD trait, without which colonies were non-viable (Otto et al. 2003). The non-viability of colonies where amoebae lacked PCD meant that, all else being equal, the PCD must have been selected for. This certainly fulfills the requirements for PCD as an adaptation. Similarly, autolysis is required for fruiting body formation in *Myxococcus xanthus* and mutants without autolysis do not develop reproductive structures (Rosenbluh and Rosenberg 1989; Wireman and Dworkin 1977).

In the *C. reinhardtii* experiments, PCD was compared to incidental death (Durand, Rashidi, and Michod 2011). Cell lysate was used as a proxy for incidental death, which again seems reasonable because the only way to ensure that no "programmed" component is part of the death phenotype is to use cell lysate. (Any damaging stimulus that causes a breach of the cell membrane/wall can cause the cell to initiate PCD pathways before the cell dies by necrosis. PCD is then not adequately controlled for.) A selection experiment between cultures with PCD and those without, or those in which death was necrotic, was not performed. However, the non-viability of cultures in which death had no programmed component means that necrotic death is selected against.

The experimental data from *E. coli, L. major, D. discoideum,* and *C. reinhardtii* provide empirical support for the historical, strict interpretation of PCD as an adaptation. A theoretical formulation of the experimental data is also possible using the Price equation and this is provided in the levels-of-selection discussion in chapter 13.

12
Ersatz PCD

THE NON-ADAPTIVE EXPLANATIONS FOR PCD

Just as there are instances when PCD is an adaptation (true PCD), there are others where it appears to be non-adaptive (ersatz PCD) (Durand and Ramsey 2019). In these cases, cell death is a side effect, an unwanted result of selection for another trait or at another level. In a sense, the difference between PCD and ersatz PCD is a mirror of Sober's "selection of" versus "selection for" distinction (Sober 1993). In ersatz PCD, selection *of* the PCD trait occurs because it is pleiotropically linked to another trait that is selected *for*. In contrast to true PCD, where death itself was selected for, in ersatz PCD a programmed mechanism leading to death is demonstrated but death itself is not selected for at the level of organization like the TA genetic module, group, kin, or population. Ersatz PCD is not an adaptation as the term is used in the previous chapter. That does not mean that at some stage it does not become an adaptation over time. Autophagy, for example, is clearly an adaptation in higher plants. It may even be that in the current scenario, death provides some degree of benefit to others, but the difference is that the benefits have not historically been selected for. This general argument has been put forward, either directly or indirectly, by several researchers (Ameisen 2002; Frade and Michaelidis 1997; Nedelcu et al. 2011; Proto, Coombs, and Mottram 2013; Ramisetty, Natarajan, and Santhosh 2015). In some cases the organism is specified, for example, the experiments by Segovia and colleagues and their argument for ersatz PCD (they do not use this terminology, which was introduced later) in *Dunaliella tertiolecta* (Segovia et al. 2003).

Autophagy and ersatz PCD

Perhaps one of the clearest examples of ersatz PCD can be made in cases where the organism dies by autophagy during starvation (for a review see Kiel 2010). Mechanistically, there are distinct autophagy genes, although it is also true that there can be molecular cross-talk between autophagy, apoptosis-like pathways, and other PCD pathways (Eisenberg-Lerner et al. 2009). It should be remembered, however, that the mechanism does not define whether it is PCD or ersatz PCD (Durand and Ramsey 2019). In the previous chapter there are cases where autophagy is an adaptation, for example in the life cycles of *Dictyostelium* and *Myxococcus*. But in many other cases autophagy is ersatz PCD. In nutrient-poor conditions, cells rely on their internal resources to survive. They recycle their intracellular components to sustain themselves until conditions improve. When there is no improvement, cells digest vital energy stores and organellar structures (mitochondria, chloroplasts, peroxisomes, Golgi bodies, endoplasmic reticulum, nuclear components) to survive the harsh conditions. By doing so they lose vital functions, become non-viable, and eventually die.

In some model organisms, the "molecular machinery for self-eating" (Yorimitsu and Klionsky 2005) has been uncovered, revealing a highly regulated biochemical pathway with several well-defined stages. These mechanisms are quite distinct from the broad PCD mechanisms sketched in phytoplankton in chapter 10. Many of the early experiments on autophagy in unicellular organisms were performed in *Saccharomyces* species and established a central role for the *Atg* genes (autophagy-related genes). At the core of most autophagy pathways is a complex of ATG proteins (Cebollero and Reggiori 2009), which are regulated by signaling molecules that sense the environment. At the onset of autophagy, a phagophore forms that engulfs cytoplasmic and organellar material. This membrane-bound structure fuses with a cytoplasmic vacuole containing hydrolases and other digestive enzymes that degrade the phagophores (now referred to as autophagic bodies). The simpler compounds (amino acids, simple lipids, and carbohydrates) are recycled back into the cytoplasm and used as an energy resource. There are other biochemical pathways and subroutines identified that contribute to autodigestion. The CVT pathway (cytoplasm-to-vacuole targeting) is similar to the process above, except that the vacuole engulfs cytoplasmic resources via smaller CVT vesicles directly (Teter and Klionsky 2000). In pexophagy, the vacuole takes up peroxisomes. Mitophagy refers to the digestion of mitochondria, and ER-phagy and ribophagy to the endoplasmic reticulum and ribosomes, respectively (reviewed in Kiel 2010). In the different manifestations of autophagy, there is considerable crosstalk, but an important feature is that in each there is a programmed component.

Besides being a form of PCD, the important evolutionary feature is that autophagy does not usually benefit others in the environment. It is probably true that the death of some individuals increases the probability of others surviving, simply because there are fewer individuals consuming resources. But there appears to be no active secretion of signals, or resources that directly benefit others (or there have not been any discovered) except in instances where the autophagy overlaps with other morphotypes like apoptosis. To the best of my knowledge, there have not been any experiments demonstrating a group-level selective advantage for autophagy, except for those with a multicellular stage like *Dictyostelium* (see below and chapter 11). In other words, in most cases autophagy is not altruism. Programmed death of the autophagy kind is a side effect of the program that has evolved as a survival mechanism. For the proponents of PCD as a non-adaptive trait, the autophagy example makes a compelling case.

The arguments against true PCD

In addition to the example of autophagy, some of the mechanistic components involved in other kinds of PCD have been the focus when counteracting the argument for PCD as an adaptation. Two of these were alluded to in previous chapters. The first general argument against true PCD is the claim that there is seemingly a significant pleiotropic component in all PCD pathways and that this may explain why a trait like death has not been removed by natural selection. The phytoplankton metacaspases discussed earlier are good examples of pleiotropy. In metazoa, caspases are frequently involved in more than one cellular function (Shalini et al. 2015) and it is expected that their distant homologs in unicellular organisms will exhibit similar pleiotropy. The argument is that, in unicellular organisms, it is much more likely that the functions of metacaspases that are unrelated to PCD are the traits that have been selected for. It is more parsimonious to expect that a trait like PCD, which is so obviously deleterious in unicellular organisms, is a by-product of another trait (for example, aerobic metabolism; Frade and Michaelidis 1997) and that the PCD trait has subsequently been co-opted in development pathways and tissue homeostasis in multicellular organisms (Huettenbrenner et al. 2003) (additional notes 12.1). There are some specific examples where this argument holds. In free-living single-celled organisms, autophagy is the mechanism by which the organism attempts to survive starvation. The enzymes responsible for breaking down intracellular resources digest vital cellular components, which eventually leads to cell death. In this scenario autophagy is the survival mechanism that is pleiotropically linked to death, which is an unwanted side effect. In *D. discoideum*, however, autophagy is required to complete the multicellular stage of the organism's life cycle (Otto et al. 2003). In this scenario, autophagy is an adaptation. In other words, a harmful trait has been co-opted to fulfill an essential developmental step.

The above example is a reasonable one for explaining how a maladaptive trait, programmed death in this case, has been co-opted for a particular developmental stage. The trait was pleiotropic in a particular context, eventually being selected for and becoming an adaptation. The issue of co-option can be expanded more generally (Nedelcu and Michod 2011). Gao and colleagues have shown that a mechanism for PCD may have evolved from a viral defense mechanism that attenuated viral reproduction (Gao et al. 2019). This provides a welcome bridge between pro-survival and pro-death mechanisms (Koonin and Krupovic 2019). In the first instance, the mechanism was initially pro-survival, protecting the cell against viral invasion. Subsequently, when death was inevitable the mechanism was co-opted at a kin / group level. PCD facilitated the survival of groups of unicellular organisms and it has been argued that selection for PCD was a prerequisite for true, complex multicellularity (Huettenbrenner et al. 2003; Iranzo et al. 2014).

It is worth noting that the arguments that programmed forms of death are non-adaptive and that they are an adaptation are not necessarily mutually exclusive. In one context, the PCD mechanism can be pro-survival and in another, pro-death. This is completely reasonable and based upon the current data, realistic. Jauzein and Erdner have made a similar argument based on their findings in *Alexandrium tamarense* (Jauzein and Erdner 2013). Their study illustrated that the executioners of PCD, like the metacaspases, can sometimes be involved in pro-survival mechanisms. Huang and colleagues demonstrated that metacaspases are implicated in PCD, aging, and acclimation (Huang et al. 2016). Teresa Mata et al. demonstrated that "type II-metacaspases are involved in cell stress but not in cell death in the unicellular green alga *Dunaliella tertiolecta*" (Teresa Mata et al. 2019) and others have provided evidence for a similar scenario in archaea (Bidle et al. 2010; Seth-Pasricha, Bidle, and Bidle 2013).

Pleiotropy explains the findings that survival and death mechanisms can sometimes overlap. But using hypothetical pleiotropy on its own to argue blanketly against adaptation is, by itself, a weak argument. First, many genes and proteins have more than one function. To begin with the assumption that, because there are two or more functions linked to a molecular pathway, one of the functions is likely to be a side effect, makes little sense. Each trait should be examined on its own merit, irrespective of the molecular underpinnings. Furthermore, if the molecular components that exhibit pleiotropy are invoked to explain PCD, how are the molecules that have only a single PCD-related function explained? Clearly, the pleiotropy argument includes some cases, but cannot be generalized to all cases of PCD.

The second, general counterargument against true PCD relates to the markers of PCD in unicellular organisms, which have fueled speculation and cast doubt on the interpretation of PCD as an adaptation (Huettenbrenner et al. 2003; Proto, Coombs, and Mottram 2013). The extension of PCD markers in multicellular organisms to unicellular ones has, as discussed in chapter 10, been problematic. An example of this is the annexin V assay for phosphatidylserine (PS) externalization. The use of this assay in multicellular organisms is appropriate, since exposure of PS on the apoptotic bodies allows phagocytic cells to engulf them via PS receptors on their surface. Notwithstanding the confusion around how the annexin V assay works in *Leishmania* and chlorophytes (chapter 10), why would PS (or any other phospholipid for that matter) externalization be important in unicellular organisms undergoing PCD? Apoptotic-like vesicles have been observed in several unicellular taxa, but their engulfment by others has been described only in *Dictyostelium*, and it is not known how this happens—whether via PS receptors or otherwise. In time, the fate of membrane-bound PCD vesicles will be uncovered, but for now PS externalization and the formation of apoptotic-like bodies remains unexplained.

In addition to the criticisms targeting the limited understanding of PCD mechanisms, there is a more general issue. It is often stated that the onus is on the adaptationists to find evidence for their claims (Hendry and Gonzalez 2008) and that the null hypothesis should be that PCD is a non-adaptive trait. Van Valen suggests that neither should have a privileged place in inference (van Valen 2009) and Gardner is critical of many of the arguments against adaptationist thinking (Gardner 2017). With respect to PCD specifically, I think there is some merit to the non-adaptationist's demand for strong supportive evidence to claim adaptation, even if I do not agree with the specific criticisms raised above (additional notes 12.2). PCD is a catastrophic event for the cell, particularly in the tradition of thinking of the organism as the target of selection. The claim that it is an adaptation is extraordinary and deserves a thorough explanation. To do so, the evolution of PCD at each of the possible levels of selection needs closer examination.

13
Programmed cell death and the levels of selection

The argument in chapter 11 that PCD is the result of natural selection is "based on the concept that unicellular life is able to organize itself into cooperating groups" (Zuppini, Andreoli, and Baldan 2007). The fitness benefits of PCD to others in the population have been demonstrated in several model organisms and, in some experiments, this was selected for. These data support the definition of PCD as an adaptation (true PCD as opposed to ersatz PCD), at least in the experimental organisms used. However, as Reece et al. indicate, the "meaning of death" (Reece et al. 2011) in most experimental studies is not always explicit, especially when its evolutionary history is not articulated. A conceptual framework for investigating PCD evolution is required.

The issue at stake for dissecting the evolutionary history of PCD is the levels-of-selection question. When are the benefits of PCD due to selection at a level of organization other than the cell, such as a group of cells? And when are the benefits to others a cross-level by-product, in which case death is of the ersatz PCD kind (Durand and Ramsey 2019)? Besides the experimental evidence discussed in chapters 11 and 12, the issue can be explored philosophically and theoretically to obtain a more general framework. The central question concerning PCD as a group-level adaptation is the relationship between the PCD trait and the fitness of groups of cells that manifest the trait. How should this relationship between PCD and selection be formulated to include the full range of the PCD trait? Okasha asks the question more generally: "When is a character-fitness covariance indicative of direct selection at the level in question, and when is it a by-product of selection at another level?" (Okasha 2006). The character-fitness covariance is the statistical relationship between a phenotype and the fitness associated with it. There are a few ways to examine this question as it applies to PCD in the unicellular world. As any student of group selection, kin selection, and multilevel selection theory appreciates, the topic is divisive and the language confusing. An understanding of the terms *units* and *levels* (at least as they are used in this book) is necessary to access the levels-of-selection arguments made by researchers in relation to the different forms of death (see table 1 and fig. 17 in chapter 8 for the evolutionary definitions of death). In the next sections, I will provide an overview of the units and levels of selection. This is also important for the synthesis presented in the next chapter. It is also essential to provide my own view on some of the issues because this directly impacts my interpretation of PCD at the different levels of selection (additional notes 13.1) and is important for the synthesis in chapter 14.

The units and levels of selection

At the outset and because of the degree to which it has percolated into the thinking of both the lay public and researchers, it should be acknowledged that Dawkins and others suggest that the terms *replicator* and *vehicle* are sufficiently illustrative of the evolutionary processes at stake. The argument is that there is really only one kind of replicating unit (the gene), which is transmitted through vehicles or phenotypes (Dawkins 2006, 1982). In this instance the levels-of-selection question is moot. Others argue that to appreciate the evolutionary processes, and the results of natural selection, requires much more than just replicators and vehicles (Lewontin 1970; Sober 1993; Huneman 2015). Hull's version of a replicator, for example, is not limited to a gene (Hull 1980). He, and many others (myself included), claim that the unit of selection can be generalized to any entity that serves as the basis for copying itself and not just the gene. In other words, the gene is not the only unit of selection, however one wishes to interpret the "many faces of the gene" (Griffiths and Neumann-Held 1999). The broad abstraction is that a level of selection is any hierarchical level of biological organization to which Darwinian principles apply (see Lewontin's criteria discussed in chapters 2 and 3). The *units* are the individuals (see chapter 6 for more about individuality) or entities that populate the hierarchical *levels*.

Some of the issues debated in multilevel selection theory are (a) whether the unit of selection is ultimately reducible to the gene, (b) whether there are discrete units at multiple levels on which natural selection can act, (c) what the levels are, and (d) how selection may work at these levels. The living world can be organized into what has become known as evolutionary transitions (Maynard Smith and Szathmáry 1995) (see chapter 6). Genes make up larger hereditary molecules like chromosomes and genomes. A prokaryote cell is the simplest extant cellular unit containing

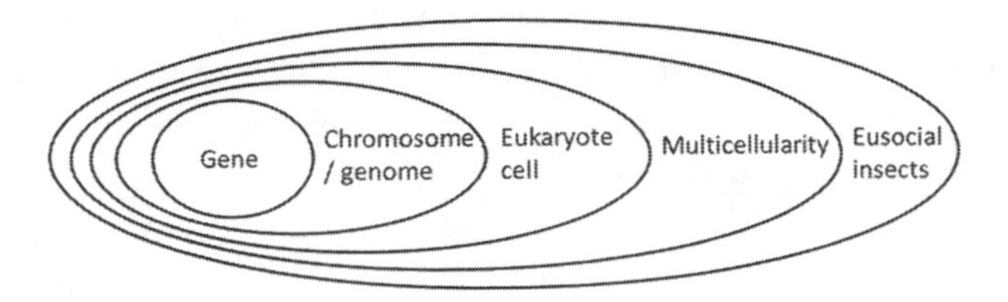

FIG. 20. The hierarchy of living systems.

the hereditary molecular material. The eukaryote cell emerged from the cooperation and functional integration of two or three kinds of prokaryote cells. Eukaryote cells make multicellular organisms, which sometimes live in obligate social communities (fig. 20). As indicated above, some authors have remained with the gene-centric view and claim that, regardless of the biological organization, the gene is ultimately the unit of selection. Others argue that different levels of organization themselves contain discrete units that are subject to natural selection. The cell, for example, may be a level of selection, but so can groups of cells be selected for. These questions strike at the heart of the debate concerning PCD evolution.

The gene-centric view of PCD

The gene-centric view has, primarily for historical reasons, been afforded a special privilege in living systems. The gene was put forward as the unit of life around which all evolutionary processes revolve and in the population genetics field, gene frequencies are used to monitor evolutionary change over time. Molecular biology entrenched the gene as the unit of change. Nevertheless, the definitions of life discussed in part 1 illustrate the limitations of this reductionist approach, and in philosophy and evolutionary biology an understanding of what exactly a gene is, has itself been a source of controversy. Bearing in mind that most, if not all, classifications in biology have exceptions, the gene can be viewed from different angles (Griffiths and Neumann-Held 1999). The so-called mobile genetic elements (also called "selfish genes"—the term is used in the strict sense, not in the general sense where every "gene" is considered selfish) are autonomous molecules and replicate as independent units, irrespective of the mechanism of replication (DNA- or RNA-mediated) (Burt and Trivers 2006). These are the transposons, retrotransposons, non-LTR retrotransposons, plasmids, and other classes of self-replicating molecules, which are sometimes collectively referred to as "replicons" (their role in the origin of life is discussed in chapters 3, 6, and 7). They have measurable fitness components (Agren 2014; Durand and Michod 2010) and there is heritable variation between individual replicons in a population. In other words, they fulfill Lewontin's criteria and evolve by natural selection (Burt and Trivers 2006). It seems entirely reasonable to conclude that, because of their autonomy (Ruiz-Mirazo and Moreno 2012; Moreno Bergareche and Mossio 2015), these versions of the gene are units of selection. The question, however, is whether this concept of the selfish gene as an individual reproducing unit can be generalized to the concept of genes that make up the hereditary material. Does the usage of the term always mean the same thing? This question matters because some researchers will argue that PCD is (only) a gene-level adaptation.

To a molecular biologist, the gene is a unit of DNA with a relatively well-defined molecular architecture that codes for a protein (sometimes more than one protein) with a particular function(s). Such genes are routinely investigated in any basic molecular biology laboratory using a reductionist approach. A cell can be manipulated genetically by knocking the gene down or out, re-introducing it, overexpressing it, cloning the gene, purifying the expressed protein, and assaying its function. This methodology reinforces the heuristic that the unit of selection can always be reduced to a gene, even though it is now known that most genes in molecular biology usually interact in some way or another with other genes (see part 1, especially chapter 2), or to be more accurate, the proteins interact (Bludau and Aebersold 2020). In the case of the molecular biologist's gene, the hereditary unit and selection unit cannot be the same thing. The vast majority of genes (excluding the selfish genes above) are not autonomous and the causal relationship between a gene and a phenotype or function is a complex one (chapters 1–3). Most extant genes may at some time in the distant past have evolved from selfish elements (Brosius 1999; Kidwell and Lisch 2000), but they no longer have any replicative autonomy or individuality (Agren 2014; Durand and Michod 2010; Kidwell and Lisch 2000) in any meaningful sense. As a discrete molecular entity, the molecular biologist's gene (except for the replicons) is an abiotic molecule and fitness is not attached to such an entity (it also fails to fulfill any of the criteria we associate with life in part 1). In addition, except for monogenic traits, there is usually some degree of epistasis (see the discussion of the genotype-phenotype map in chapter 2).

To some evolutionists, the gene is defined as a stretch of DNA that is transmitted intact from parent to offspring (a DNA sequence that does not undergo recombination during meiosis). This interpretation, championed by Dawkins (Dawkins 1982, 2006), argues that even though there are interactions (epistasis) the gene is still a discrete heritable unit, just not defined in the same way as molecular biologists use the term. The argument that the gene is a discrete heritable unit must be true: after all, that is how it is defined (Dawkins argues that the discrete heritable nature is necessary for the gene-centric view; see Dawkins 1982). The issue, however, is that there is no reason to equate the hereditary unit with the selection unit in this version of the gene. The lack of autonomy or individuality associated with this version of the gene as well as the molecular biologist's version, remains problematic. It can be argued that a genome-centric view, proposed by Heng, will harmonize the hereditary unit with the selection unit (Heng 2009) because the genome is both hereditary and autonomous. Theoretically at least, the genome is a much better approximation (certainly better than the gene) for a discrete, replicatively autonomous individual. This is helpful for prokaryote genomes. But in sexually reproducing eukaryotes, the genome undergoes recombination during meiosis, which renders it unhelpful as the "intact" hereditary unit or for tracking genetic changes in populations over multiple generations.

Considering the conceptual flaw (equating the hereditary unit with the selection unit) and limitations of the gene-centric view of life, why is it so often used by biologists in evolutionary arguments? The reasons are theoretical, practical, and historical. Genes have been used theoretically in population genetics since the works of Wright, Fisher, and Haldane (Kempthorne 1983). They are a good proxy for monitoring or measuring evolutionary change, even if they do not stand up to scrutiny as the unit of selection. Molecular biologists have entrenched this practice because of the interest in dissecting out the activities of specific genes and proteins. Historically, the gene-centric view also benefited from the broad public appeal of the idea of a selfish gene and became the default view, especially following the initial conceptual shortcomings of group selection theory.

Which interpretation of the gene should be used to examine gene-level selection in PCD evolution?

PCD researchers have usually not stated explicitly which interpretation of the "gene" they use. I think it is most appropriate to use only the strict interpretation of the gene as an autonomous unit such as a replicon (MGEs, plasmid-encoded TAs, etc.) described above. The first reason is that PCD genetic modules are in some cases functionally equivalent to replicons. The second reason is more general. The other versions of the gene discussed above (not the replicon interpretation) are not units of selection. Non-autonomous genes have no direct attachment to fitness. On their own they have neither viability nor reproduction and for this reason alone,

I would argue that genes (except for replicons) are never the units of selection. They are hereditary units, but that is something quite different.

In the case of replicons, PCD can be a gene-level adaptation. Cells that harbor the autonomous PCD TA-encoded genetic module are viable, but those that lose the module by whatever means die. PCD is directly linked to the *function* of the TA module, which results in selection for or against it. Without PCD, the TA module is lost. Thus, PCD is maintained by gene-level selection.

This interpretation can be at odds with the views of others. For example, Hazan, Engelberg-Kulka, and Kolodkin-Gal seem to attribute PCD induced by TAs like *MazEF* to group- or population-level selection (Hazan and Engelberg-Kulka 2004; Kolodkin-Gal et al. 2007) (discussed in chapter 11). I would suggest that only in cases where the *MazEF* mechanism includes a functional dependence on group-level traits like those described by Kolodkin-Gal et al. (2007) should group-level selection be invoked, although this should still be tested explicitly. Ramisetty and others do not accept TA mechanisms as examples of true PCD where the PCD trait itself has been selected for (chapter 8) (Ramisetty, Natarajan, and Santhosh 2015; Ramisetty and Santhosh 2017). In this instance it seems to me that the issue is really about how PCD is defined. I would agree that in these cases PCD is not selected for but I claim that this is not for the reasons that Ramisetty and colleagues propose. Instead, I suggest that it is gene-level selection with cell death as a side-effect.

Ameisen views TA mechanisms as addiction molecules, without the need to invoke higher levels of selection (Ameisen 2002). He suggests that cells are "addicted" to the TA genes since those that harbor them die via PCD if the genetic module is lost or loses function. I agree with this interpretation, but only when the TA module is not autonomous, such as when it is chromosomally encoded. In these situations, the TA module is not autonomous and not an individual in any meaningful sense. The cell's fate is entirely evolutionarily aligned with the TA module and in that sense the cell is addicted to it. But chromosomally encoded TA genes cannot be a gene-level adaptation because the word *gene* is not used in the autonomous, MGE sense. Chromosomally encoded PCD genes that are not autonomous are simply genes of the molecular biologist's kind, to which the genome has become "addicted." In the same way, the bacterial genome can also be considered addicted to any gene that is essential for survival. However, when the TA module is autonomous (MGEs, plasmids, replicons, etc.), then the gene-level selection argument holds and PCD is a byproduct, the result of cross-level selection.

Cell-level selection and PCD

In some bacteria such as *Bacillus* and *Myxococcus* spp., PCD is part of the organism's development. In *B. subtilis* (Smith and Foster 1995) and *B. anthracis* (Chandramohan et al. 2009), for example, mother cell lysis occurs, and a spore is released. The death of the mother cell involves a PCD pathway and PCD is selected for because sporulation is a developmental component in the organism's life cycle. Without PCD, sporulation is not possible and the organism is non-viable. In these cases, PCD can be interpreted as cell-level selection. The term PCD, however, is clearly counterintuitive when used in this context, even if some of the molecular mechanisms overlap with other forms of PCD. The bacterial genome does not die in that it ceases to exist; instead, it survives in a new, different kind of cell. As discussed in chapter 8, the terminology can be very misleading, and I think that this is one of those cases where the term PCD is truly problematic. Clearly, if a single-celled organism dies, this cannot be a cell-level adaptation. Nevertheless, PCD is sometimes used to describe the process of sporulation in some bacteria. It should be remembered, however, that in these instances it describes a developmental stage as opposed to the termination of the organism.

Kinship, kin selection, and PCD

Kin selection theory (Gardner, West, and Wild 2011; Michod 1982; Maynard Smith 1964) emerged from the centrality of the gene in population genetics to explain seemingly unexpected traits like altruism. Hamilton's insights are summed up by the statement that costly behaviors can evolve if the beneficiaries are genetic relatives and the cost/benefit ratio is less than the degree of relatedness, so $(c/b) < r$ (Hamilton 1964). His idea of inclusive fitness is the essence of kin selection theory advanced by Maynard Smith (Maynard Smith 1964). From the point of view of inheritance, kin selection makes complete sense. It provides a general framework for explaining sociobiological phenomena like cooperation and altruism (altruistic behavior in one individual would readily evolve if the additive benefits in relatives outweigh the costs to the actor) and there is extensive empirical evidence from a range of organisms to support kin selection models (Dugatkin 1997; Lyon and Eadie 2000; Strassmann et al. 2011) (see additional notes 13.2 for the difference between cooperation and altruism). There are, however, explanatory limitations (Birch 2014; Birch and Okasha 2015), although Queller (Queller 2016) suggests that criticisms of the theory are largely the result of historical idiosyncrasies and the nature of developments in the field rather than any fundamental issue.

It is worth noting that kin selection theory is rooted in the gene-centric view. Whatever the measure of kinship (alleles, polymorphic sites, whole genomes, pedigree, etc.) (Blouin 2003), relatedness at the genetic level is the central ingredient used to explain costly phenotypes in social behavior. Relatedness, when measured by genes, is typically defined as the average proportion of alleles of an individual that are identical by descent to those of another. In other words, this equates to the probability that two individuals share the same allele, derived from the same parent at a particular locus. Changes in gene frequency are useful for monitoring evolution because they (usually) reflect the process of natural selection accurately. However, because kin selection theory is rooted in the gene-centric view, the measures of kinship also conflate the heritable unit with the unit of selection. This is an important caveat. Some authors are also critical of the causal structure of kin selection theory and Hamilton's rule (Nowak, Tarnita, and Wilson 2010) and are certainly critical of its universal implementation (Nowak et al. 2017). A host of authors disagree with the criticisms (see, for example, Abbot et al. 2011; Herre and Wcislo 2011) but the debate will persist. I would agree that in many instances, kinship does *reflect* (not necessarily capture mathematically) the biological underpinnings of traits like altruism. However, I would also add that this is not always the case even if the beneficiaries of the altruistic act are typically genetic relatives (Birch 2014). The reason for this is in part because genes (except for replicons) are not, in my view and some researchers would agree, the units of selection. The methods for measuring kinship are also not always agreed upon (Birch 2014; Birch and Okasha 2015; Goodnight 2013a; Nowak, Tarnita, and Wilson 2010). The important issue, however, is the more fundamental question concerning non-genetic inheritance and how this impacts the evolutionary process (Bonduriansky and Day 2018; Laland 2015; Laland et al. 2015; Sultan 2015). It should be remembered that there are many other ways to examine PCD, beyond the traditional levels-of-selection approach. Examples are evolution by niche construction and ecological inheritance (these are mentioned briefly in the postface and not included in this book because they have hardly been explored in PCD research). Kin selection theory on its own does not capture all the evolutionary processes observed, including altruism. Nevertheless, I would agree that despite the philosophical and empirical objections that stem from the limitations of the gene-centric view, kin selection theory is a powerful framework for investigating many of the problems in sociobiology (Michod 1997, 1999; Michod and Roze 1997; Queller 1992a, 2000, 2016).

Kin selection theory certainly informs our understanding of PCD (Vostinar, Goldsby, and Ofria 2019). As Reece and colleagues have demonstrated, "relatedness regulates death" (Reece et al. 2011). In a clonal population of unicellular organisms, where r = 1, if the death of one cell by PCD facilitates the survival of more than one of its clonal relatives, there is a net inclusive fitness benefit and PCD can evolve by natural selection. This is, after all, what explains PCD in multicellular organisms. In contrast, when a cell dies without the PCD mechanism and there is no increased survival in relatives, this is a disadvantage, especially when death without PCD is harmful to others (Durand, Rashidi, and Michod 2011). Reece et al. have modeled the scenarios under which natural selection favors the evolution of a PCD strategy in *Plasmodium* and illustrated how rates of PCD vary depending on relatedness (Reece et al. 2011). Vostinar and colleagues have provided more general models (Vostinar, Goldsby, and Ofria 2019). Another mathematical model by Iranzo et al. demonstrated the evolution of PCD (together with aggregation; see chapter 16) is an optimal strategy in response to viral invasion of the population (Iranzo et al. 2014). Theoretical studies like these and the limited number of empirical findings are drawn upon to explain PCD evolution as an example of kin selection.

Group selection and PCD

Group selection theory has been included, sometimes unintentionally, in the interpretations of empirical data for PCD. However, to invoke group selection theory in the explanations for PCD, especially where it is used in the synthesis in the next chapter, it is important to detail how the term is used here (Durand and Ramsey 2019). In the early historical stages of the development of group selection theory the "good of the group" argument was often invoked to explain behaviors that were difficult to assign to individuals in the group. This line of thinking was justifiably countered by Williams (Williams 1966), who demonstrated that in most instances, the phenomena that group selection was used to explain could be understood more parsimoniously. Maynard Smith was also critical of naive group selection because of the vulnerability of the group to disintegration by selfish individuals, arguing that kin selection provided a more robust explanation for traits like altruism (Maynard Smith 1964). Group selection as a general process in evolution was largely dismissed as artifactual in the early stages of its theoretical development and group adaptations, it was suggested, were likely non-existent. The claim was that group benefits were common, but not the result of group selection; rather they were a side effect of selection or other processes at the level of individuals. This is the issue (as well as the kin selection theory above) that strikes at the heart of the PCD evolution debate.

In many evolution circles, the anti-group selection sentiment is still very much intact and group-level selection is excluded as an explanation for PCD (and everything else). The complete rejection of group selection (which dates back to the 1960s) as a valid concept and its subsequent rejection in understanding PCD, however, is in my view a mistake (and again many other researchers have the same view). A return to the uncritical "good of the group" argument is obviously not called for, but many researchers (for example Borrello 2005; Sober 1993; Sober and Wilson 1994; Eldakar and Wilson 2011) successfully resuscitated the discipline by reformulating what is meant by group selection. The theory is now part of mainstream evolutionary thinking. In a purely mathematical and functional sense, group and kin selection can be considered equivalent (Lehmann et al. 2007; Marshall 2011), although conceptually and mechanistically they require different causal explanations (Okasha 2016). In real life examples, they also need not be mutually exclusive (Pievani 2014). The concepts of kin and group selection are sometimes used casually to mean the same thing, but their causal structures are important for understanding PCD evolution and there are good reasons to keep them separate. Wilson's "trait group" model (Wilson 1975) elegantly demonstrated that altruism, which PCD clearly is when it is an adaptation, is not necessarily limited to kin. A potential example of this is the evolutionary transition that gave rise to the eukaryote cell (Margulis 1981; Michod and Nedelcu 2004). Evolutionarily distant prokaryote taxa may have cooperated to form a eukaryote cell (there are other hypotheses for eukaryogenesis discussed further in chapter 16 and see also Zachar and Szathmáry 2017). The mitochondrion is central to PCD in unicellular eukaryotes, but clearly relatedness was not a prerequisite for the evolution of altruistic death in eukaryogenesis.

One of the more surprising discoveries in unicellular PCD research was the observation that "altruism [in reference to PCD] can evolve when relatedness is low" (Refardt, Bergmiller, and Kümmerli 2013). This was a significant advance in the field, demonstrated in a selection experiment where one population of *E. coli* with PCD outcompeted one without PCD, if the starting frequency of the PCD trait was sufficiently high. In this instance, the causal structure for PCD evolution was not clear. Relatedness was not quantified but the experimental setup ensured that it was low. The frequency of the costly trait in the population should therefore have also been relatively low. The empirical data, however, indicated that the starting frequency was unexpectedly high for PCD to evolve. In these experiments the quantification of "r" and subsequent calculations should uncover these relationships, but it appears that the essential ingredient here was not relatedness. This points more toward a group selection causal structure. Experimental evidence for group selection had been demonstrated much earlier, notably by Wade, Goodnight, and others (for example, Goodnight 1990b, 1990a; Wade 1977; Wade and Goodnight 1991) and the mathematical models modified accordingly (Wade 1978), but the *E. coli* experiments contextualize the kin / group selection question to PCD.

It should be cautioned that just because there is little or no relatedness in communities where PCD is common, group selection is not always the appropriate approach. This is especially the case where PCD in one taxon benefits other taxa. PCD in *D. salina*, for example, benefits not only others of the same strain but also a co-occurring archaebacterium (Orellana et al. 2013). Examples like these are well known in phytoplankton communities (Bidle 2015, 2016; Bidle and Falkowski 2004; Franklin, Brussaard, and Berges 2006) and group selection is not the preferred approach to explain the benefits of PCD to other taxa in these instances. The conditions for group selection to act are unlikely to be present and there are other approaches like reciprocal altruism that are much more convincing. As Ramsey and Brandon, West et al. (West, Griffin, and Gardner 2007), Birch (Birch 2017), and others have clarified, "Reciprocal altruism is not a kind of group selection" (Ramsey and Brandon 2011) because of the causal structure of selection in these instances. As indicated in the section on kin selection, discussing all the possible explanatory frameworks like reciprocity is beyond the scope of this book (the postface alludes to some of the other potential explanatory frameworks).

Population- and species-level selection in PCD

The nature of the evolution of species by punctuated equilibrium (Gould and Eldredge 1977) and later by species-level selection has undergone several revisions (Gould 2002) since the concepts were first introduced. As with so much else in multilevel selection theory, the theory and data support aspects of species selection, depending on how the concept is defined (Jablonski 2008). The challenge in species-level selection, or any other level of selection for that matter, is to determine whether the trait-fitness covariance relationship is causal or a cross-level by-product (Okasha 2006). Species-level selection occurs when trait and fitness are causally related. In its strictest form the case for species selection can be made when the trait in question is emergent at the level of the species and has a bearing on species fitness (fitness is measured by speciation or extinction rates). There must also be screening off from selection at other levels. A frequently cited example of a relevant emergent trait in species-level selection is population structure. The character of population structure cannot be reduced to a single organism in the population; of course, at some level it depends on the individuals, but the population structure relies on there being a population.

PCD evolution with respect to species-level selection has not been dealt with in the literature, but it is nevertheless something that should be considered. The finding that PCD can have differential fitness effects depending on the species (Durand et al. 2014) may mean at least two things. First, if PCD emerged after the speciation events (this seems unlikely because PCD appears to be an ancient phenomenon) then the differential fitness effects may simply reflect interspecies differences. Second, there is also the possibility that varieties or different populations of the same species were differentially impacted by PCD, leading to greater divergence and eventually speciation. Species-level selection, however, is mentioned here mostly for the sake of completeness. Except for a cursory comment later, PCD evolution in the context of species-level selection is not discussed further.

Formulation of PCD and the levels of selection

The key question that arises when interpreting the data concerning the benefits of PCD is this: how do we know whether the fitness advantages to others have evolved due to selection or whether the benefit is a cross-level by-product? This question can be operationalized in a few ways.

The nature of selection at the different levels of organization can be formulated with Damuth and Heisler's multilevel selection approach (Damuth and Heisler 1988; Heisler and Damuth 1987), which has been applied to several other levels-of-selection questions (see for example Damuth and Heisler 1988; Frank 1998; Michod 1999; Queller 1992b; Sober and Wilson 1998). Theoretically at least, the trait-fitness covariance at different levels can be partitioned with the Price equation (Price 1970, 1972). Price's formulation is a statistical one and the issue of causality still needs to be demonstrated either empirically or with other mathematical models, but it provides a relatively simple conceptual framework for generalizing the

PCD question. It is used in Okasha's general multilevel selection (MLS) framework (Okasha 2006) and can be used to formulate PCD evolution (Durand and Ramsey 2019). Before applying the Price equation to PCD, it should be noted again (see chapter 8) that PCD is not a discrete all-or-nothing variable (Durand and Ramsey 2019). It is a continuous trait. The environmental triggers that activate the PCD pathway can lead to a range of outcomes including anastasis (anastasis has generally been examined in multicellular organisms, but a similar scenario plays out in the unicellular world) (Sun and Montell 2017), cell cycle arrest (Helms et al. 2006; Torgler et al. 1997), dormancy or encystation (Khan, Iqbal, and Siddiqui 2015), and differentiation (Cornillon et al. 1994), as well as PCD (chapter 8). Cell viability varies in each of these outcomes but does not diminish to zero (except of course at the final stage of PCD), since reproductive potential remains to varying degrees.

There are many decompositions of the Price equation (Luque 2017). For PCD, the experimental designs that investigated potential group-level effects are appropriate for MLS1 (groups are aggregates of individuals and the individuals are the focal units) as opposed to MLS2 (where the group is the focal unit). The reduced version of the Price equation is

$$\overline{w}\Delta\overline{z} = Cov(w_i,z_i) \qquad \text{(Eq. 1)}$$

where $\overline{w}$ is the average individual fitness, $\Delta\overline{z}$ is the change in the average of the character trait (in this case PCD), from one generation to the next and $Cov(w_i,z_i)$ is the covariance between fitness and character trait for the ith individual. The overall character-fitness covariance of the entire population comprises two parts: the covariance between groups and the average (or expected) covariance within groups,

$$Cov(w_i,z_i) = Cov(W,Z) + E(Cov(w,z)) \qquad \text{(Eq. 2)}$$

which allows the product of the average fitness and average change in character trait of the population to be written as

$$\overline{w}\Delta\overline{z}=\text{Cov}(\text{W, Z}) + \text{E}(\text{Cov}(\text{w,z})) \qquad \text{(Eq. 3)}$$

For any individual, the mean fitness and the change in the mean of the character depend on the covariance at the level of the group (first term) and at the level of individuals in the group (second term). The question for PCD as a group-level adaptation hinges on knowing whether both terms in the Price equation are necessary to explain the observed data. In other words, can

$$\overline{w}\Delta\overline{z}$$

be explained by the second term alone (covariance at the level of the individual cell), or is the first term (covariance at the group level) also required to explain the empirical observations?

There are two points worth noting before interpreting the empirical data with Eq. 3. First, the assumption is made that there is no transmission bias in PCD, and that the trait is transmitted faithfully from parent to offspring. In other words, the assumption is that the evolutionary change is due to natural selection alone (additional notes 13.3). There are different decompositions of the equation that separate out transmission bias and natural selection (Luque 2017), but these include additional terms for which there are no empirical data for PCD. More importantly, the assumption of no transmission bias is actually a worst-case scenario because individuals with the PCD trait die or have lower viability or reproductive potential. If there is any transmission bias at the individual level, it diminishes the evolutionary response rather than enhancing it, since the trait is not passed faithfully from parent to offspring. Second, it should also be remembered that the character "z" in question, PCD, is treated as a continuous trait (see above). The loss of viability is graded and non-discrete. At one end of the spectrum, PCD may simply be a transient hiatus in cell cycle progression. At the other end, there is the immediate implementation of the genetic program for death. Between these two extremes, there are "degrees of death" like prolonged arrest in the cell cycle, senescence or some other loss of viability, encystation and spore formation, and degrees of autophagy (Thomas et al. 2003).

The experiments with *E. coli* (Refardt, Bergmiller, and Kümmerli 2013), *L. major* (van Zandbergen et al. 2006), *D. salina* (Orellana et al. 2013), *S. cerevisiae* (Herker et al. 2004; Fabrizio et al. 2004), and *C. reinhardtii* (Durand et al. 2014; Durand, Rashidi, and Michod 2011) are some of those that are accessible for interpretation with the Price equation. Calculating the covariance was not the aim in these experiments, but what is clear from the data, and indeed intuitively obvious, is that fitness and PCD have an inverse relationship. As the PCD pathway is implemented, the cell gradually dies and fitness decreases. The second term in Eq. 3, $E(Cov(w,z))$, is negative. The experimental results showed that in cultures where PCD occurred, the remaining individuals produced more offspring. In Eq. 3, the left-hand side is positive, since the change in PCD ($\Delta\overline{z}$) (measured directly in the *E. coli* experiments) is positive. The second term on the right-hand side, the individual character-fitness covariance, is negative. It can be concluded, therefore, that the term $Cov(W,Z)$ must be positive. Interpreting the empirical data with the Price equation, thus, suggests that at the group level, PCD and fitness can covary positively. In other words, the data do fit with selection at the group level.

In the MLS1 situation, the second term in the Price equation must be included because the individuals are the focal units. However, it remains to be seen whether in all situations, a non-zero value can be assigned to group-level covariance (the first term). A feature of the Price approach to MLS1 is that the group and individual covariances are neatly compartmentalized. In many instances, however, this may not be appropriate for the experimental designs and it is possible that the absence of "screening off" (Brandon 1990; Brandon and Carson 1996) between levels biases the results. There is a second methodological approach that may help with this potential bias.

It is possible to contextualize PCD data to the group to examine this independently of individual level selection. Heisler and Damuth (Heisler and Damuth 1987) propose this general contextual analysis approach by apportioning an individual's fitness (w) to both the individual's character (z) as well as the group character (Z) (Eq. 3). In this formulation, the regression coefficients β_1 and β_2 measure the effects of individual character and group character on individual fitness, respectively. The issue here is that the condition $\beta_2 \neq 0$ must be satisfied to conclude that there are true group-level effects on an individual's fitness.

$$w = \beta_1 z + \beta_2 Z + e \qquad \text{(Eq. 4)}$$

Applying the same logic above, the first term decreases average individual fitness, while the second term increases it. The experimental results indicated that mean fitness increased in the populations, which means that the second term, which reflects the group effect on individual fitness, must be positive. Formulating PCD and its fitness effects in this way leads to the same conclusion, that there are group-level effects that increase an individual's fitness.

The support that the Price equation and contextual analysis bring to the argument that PCD is a group-level adaptation should always be treated with caution. Both approaches are general formulations and are simply ways of operationalizing changes in fitness and traits in the population. While they reveal that the data indicate that there are indeed group-level effects, they are, of course, abstract. Experimental designs that explicitly examine whether these effects are the result of an evolutionary response are still required (see Frank 2012 for a review of the Price equation). Nevertheless, they are helpful in the sense that they can be used to examine PCD as a continuous trait in a range of future contexts. They not only capture the empirical findings but bring a general formulation to the synthesis of the origin of PCD.

14

A synthesis for the origin of programmed cell death

The philosophical, theoretical, genomic, and empirical data in part 2 permit a broad synthesis for how PCD emerged in the unicellular world. Beginning with the simplest form of PCD and following a parsimonious "bottom up" trajectory, eight major steps are identified. Some of these would have overlapped and they were certainly not as discrete as portrayed here. But our current knowledge suggests that these are the key innovations that recapitulate the path leading to the manifestations of PCD in extant unicellular organisms seen today. Not all elements of PCD are included in this broad overview. Sporulation and death of the mother cell, for example, may involve PCD genes, but these are more specialized outcomes of the PCD pathways. Furthermore, at each stage there is diversification and many different versions of the same process are found in different taxa and ecological contexts, but these varieties are not conceptually different. The synthesis includes only true PCD as defined in chapter 8 and not ersatz PCD, where the genetic program for death is a by-product and death itself has not been selected for (see Durand and Ramsey 2019 for the definitions of death). The major steps in PCD evolution are referred to as such because they require evolutionary innovations.

The evolution of PCD in the unicellular world in eight major steps

STEP 1: GENE-LEVEL SELECTION OF AUTONOMOUS REPLICONS (FIG. 21)

The simplicity of the toxin-antitoxin (TA) mechanism and its presence in both archaea and bacteria (Yamaguchi, Park, and Inouye 2011) suggest an ancient origin for this mechanism. Of course, it is possible that the TA mechanism emerged later and subsequently infected all prokaryote lineages. However, its relative simplicity and ubiquity suggests that it, or

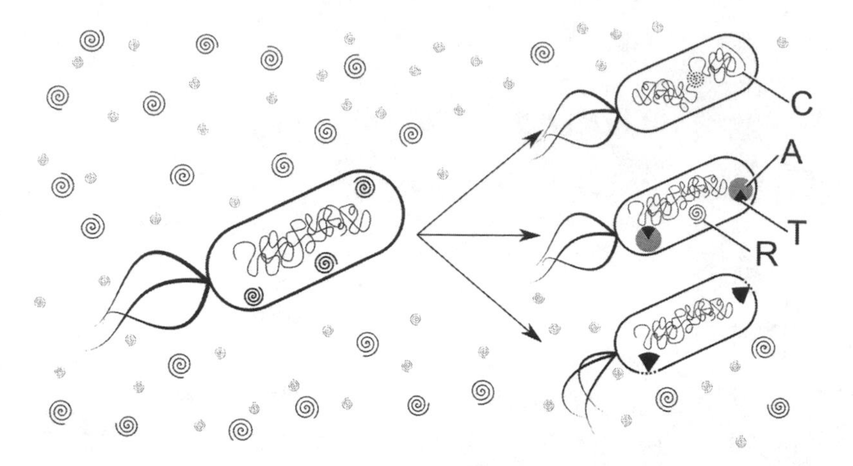

FIG. 21. PCD and gene-level selection. Autonomous genes, also known as replicons (R), that code for the TA module invest in their own survival. The labile antitoxin (A) inhibits the activity of the stable toxin (T). In cells where the replicon is lost, the T outlives the A and induces death in the cell. In this way, replicons ensure their own survival. In some instances, the replicon can translocate to the bacterial chromosome (C). In this instance the replicon may lose its autonomy. Its replication is under the control of the chromosome and its evolutionary fate is aligned with that of the bacterium.

a version of it, was possibly one of the earliest genetically encoded programs that caused cell death. TA genes can be encoded by autonomous replicons. (The somewhat abstract term *replicon* is used here for convenience to indicate that the genetic module is autonomous. Its replication is independent and uncoupled from the prokaryote genome and is a *selfish gene* in the strictest usage of the term [see chapter 13].) Replicons may take the form of MGEs on extra-chromosomal plasmids (see chapter 8 for a discussion of TAs, replicons, and the different terminologies). In these situations, the fate of the MGE and the cell are uncoupled, and cell death is a gene-level selection mechanism for ensuring the persistence of the TA-encoding replicon. Cells with the MGE are protected by the labile antitoxin, while those that lose the MGE die due to the more stable toxin that persists intracellularly after cell division. However, if the MGE translocates to the prokaryote genome, it may lose its replicative autonomy over time and the evolutionary fate of the cell and the MGE are then aligned. At this stage, death impacts both the cell and the TA genetic module and another evolutionary step is required for PCD to be selected for.

FIG. 22. Protection against harmful, unregulated death. The genetic program for death (PCD) is chromosomally encoded. (A) Activation of PCD leads to the dissolution or modification of toxic metabolites and enzymes stored in bacterial microcompartments (BMCs). Disintegration of the cell, therefore, does not harm others. (B) In cells without a PCD program, the cell lyses and spills its contents, including the toxic waste and enzymes sequestered in BMCs, into the environment, with the potential for harming or killing others in the vicinity.

STEP 2: PROTECTION AGAINST UNREGULATED CELL DEATH (FIG. 22)

From the cell's perspective, once a programmed mechanism for cell death is hardwired in the prokaryote genome, a second innovation is required. The mechanism may be related to TAs translocating to the cell's genomes, but the mechanism can also be unrelated and arise de novo. Either way, if this phenotype is to be maintained by natural selection (true PCD) and not merely non-adaptive means (ersatz PCD) a second step in the evolution of PCD is necessary. This came about because, like all cells, prokaryotes contain toxic metabolites, oxidizing compounds, and enzymes that are damaging unless they are sequestered in intracellular microcompartments (for example the BMCs, bacterial microcompartments) or inactivated (Koonin, Makarova, and Wolf 2017; Saier 2013). Liberation of these cellular components is harmful to other cells in the vicinity. This has been demonstrated in the eukaryote *Chlamydomonas* (Durand et al. 2014; Durand, Rashidi, and Michod 2011), but it is extended here to prokaryotes.

It seems reasonable, and parsimonious, to assume that PCD first emerged because it protected relatives against the harmful effects of unregulated death. The release of toxic materials would have been damaging to genetic relatives, and kin selection is a likely explanatory framework that led to neutralization of harmful unregulated death. The mechanisms that are responsible for this early form of PCD can involve fairly sophisticated protein pathways. The genes coding for these proteins are chromosomally encoded (they are not autonomous like the TAs) and are phylogenetically conserved (Aravind, Dixit, and Koonin 1999; Nedelcu 2009). These forms of PCD are common in bacteria (Koonin and Aravind 2002) but appear to be less common or mechanistically simpler in the archaea. This suggests that they may have emerged either in the ancestor of bacteria and archaea with subsequent simplification of the mechanism in archaea, or they could have emerged near the bacteria-archaea split with greater sophistication developing in the bacteria. There is some evidence that archaea can undergo PCD (Bidle et al. 2010), but uncovering this is still at the very earliest stage, and the more complex mechanisms that have persisted in eukaryotes appear to have been inherited predominantly from the true bacteria (Koonin and Aravind 2002). Protecting against the harmful effects of PCD may have coevolved with the beneficial effects of PCD in steps 3 and 4. The PCD mechanisms in these steps are much more complex and it seems more likely that they emerged after those that neutralize the harmful effects of unregulated death.

STEP 3: PROTECTION AGAINST VIRAL PARASITES (FIG. 23)

The innovation that led to PCD actively benefiting relatives would have emerged after the more passive mechanisms for simply neutralizing unregulated death. It is possible, at least in some organisms, that the mechanisms for PCD as an adaptation may have been co-opted from pro-life pathways. Gao et al. discovered that PCD pathways could have emerged from pro-life pathways that attenuated viral infection (Gao et al. 2019). In situations where cell death was inevitable, the emergence of PCD provided kin / group-level advantages. The frequency of the PCD trait needs to be relatively high in the population before the benefits are realized (Refardt, Bergmiller, and Kümmerli 2013), suggesting that PCD was being maintained as a stable trait in populations before this step was possible. The evolution of PCD (and aggregation; see chapter 16) is the optimal evolutionary strategy under conditions of high viral load (Iranzo et al. 2014). The benefits of PCD at this stage of its evolution are two-fold. The neutralization of unregulated death described in step 2 benefits others passively. The implementation of altruistic death also benefits others in the population because infected cells commit suicide before the infecting virus replicates and lyzes the cell, liberating new virions that go on to infect and lyze others. This second advantage of altruistic death may extend to any bacteria in the vicinity that are vulnerable to infection by the virus. Kin selection alone can explain adaptive PCD at this stage (Vostinar, Goldsby, and Ofria 2019). At the same time, however, relatedness between

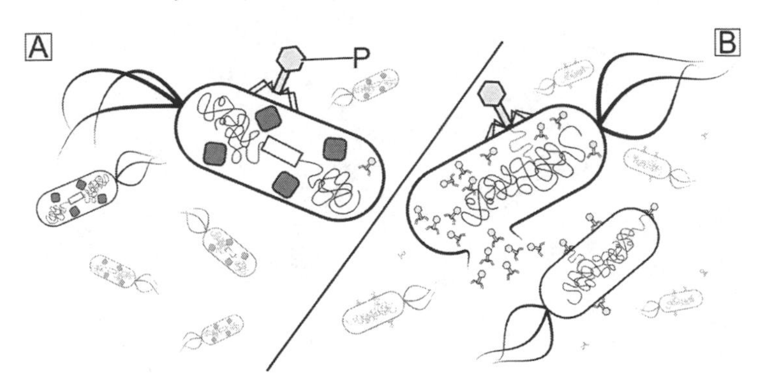

FIG. 23. Group-level protection against viral parasites. (A) Bacteria with a PCD program undergo altruistic death before the infecting bacteriophage (P) replicates. (B) In bacteria without PCD, the cell lyses, liberating virions that infect others.

individuals in the vulnerable population can sometimes be very low and PCD is still advantageous (Refardt, Bergmiller, and Kümmerli 2013). It is expected, therefore, that this step in PCD evolution is driven by either kin selection or group selection (depending on the cost-benefit ratio and the degree of relatedness, see chapter 13).

STEP 4: RESOURCE SHARING AND SIGNALING IN COOPERATIVE GROUPS (FIG. 24)

Step 4 is an alternate strategy to step 3 in PCD evolution. In some lineages, step 3 may have emerged, in others step 4, and in others still the two strategies may have coevolved. In step 4, the benefits of PCD are unrelated to viral invasion. Instead, PCD presented new ways for individuals to communicate and share resources. Communication may take the form of signaling molecules that are released into the environment during the PCD process and that convey information to others about the environmental conditions (for example, quorum-sensing molecules). Metabolites can be detoxified and released to provide nutrients to others (public goods). For these mechanisms to emerge, new pathways were required. The finding that PCD proteins are sometimes involved in more than one cellular process (for example, PCD can be connected metabolically to lipid biosynthesis; Sathe et al. 2019) suggests that some of the molecular components could have arisen by co-option of genes or exaptation. In addition, there may have been gene duplication and neofunctionalization. Alternatively, PCD genes may have arisen de novo and subsequently been co-opted into unrelated pathways.

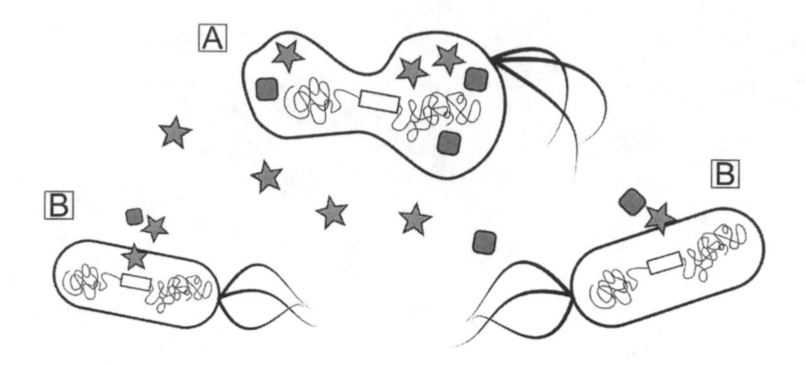

FIG. 24. Resource sharing and signaling in cooperative groups. When undergoing PCD, (A) cells liberate resources, detoxified metabolites, or signaling molecules that are (B) taken up by others. This imparts a fitness benefit to the recipients.

STEP 5: MULTICELLULAR-LIKE BEHAVIOR IN PROKARYOTES (FIG. 25)

PCD is an essential survival component in some bacterial communities and has been implicated in the multicellular-like behavior observed in prokaryotes. The multicellular-like features that form part of bacterial communities include sharing resources, communication, detoxification of the environment, differentiation and division of labor, and the regulation of population growth and carrying capacity (Bayles 2007, 2014; Lyon 2007; O'Malley and Dupré 2007; Rice and Bayles 2008; Nadell, Xavier, and Foster 2009). The mechanistic basis for PCD-related communication in prokaryotes with multicellular behavior is being uncovered (for example, Kolodkin-Gal et al. 2007; Dyrka et al. 2020). Biofilms are good examples of these complex, multicellular-like community structures. Depending on the cost-benefit ratio and the degree of relatedness between individuals as well as population structures, kin or group selection or both are the explanatory framework.

STEP 6: THE EUKARYOTE CELL (FIG. 26)

The evolution of the eukaryote cell was a major evolutionary transition in individuality (Michod and Nedelcu 2004; Blackstone 2016; Maynard Smith and Szathmáry 1995) and PCD was a key component of this process (see chapter 16). There were significant advantages to cooperation between completely unrelated taxa, such as an increase in cell size, storage

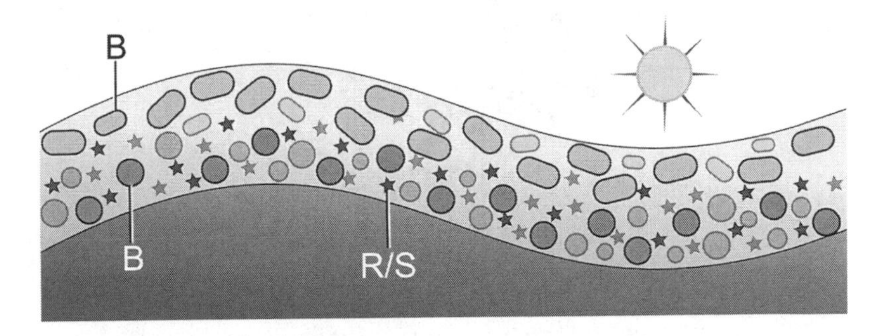

FIG. 25. Multicellular-like behavior in prokaryotes. Bacteria (B) sometimes live in complex communities where individuals of the same clone may exhibit different phenotypes and functions (division of labor or differentiation indicated by different shapes in the biofilm). Some individuals may undergo PCD, releasing resources and signaling molecules (R/S) that are used by others.

of resources, more sophisticated mechanisms of communication between cells, avoidance of predation, endocytosis of extracellular materials, and access to more efficient mechanisms of respiration and energy production (Blackstone 2013). However, the individuals that cooperated to form a eukaryote cell (archaebacteria, proteobacteria, and later cyanobacteria) were unrelated and there was, and still is, genetic conflict between genomes within the eukaryote cell (Blackstone 2013; Michod and Nedelcu 2004; Burt and Trivers 2006). The migration of genes between cellular compartments aligned the evolutionary interests of the different components to some degree, but PCD remains a mechanism for mediating genetic conflict between the three cellular genomes (nucleus, mitochondrion, and chloroplast) that evolved from their prokaryote ancestors (Blackstone and Green 1999). PCD ensures that the different cellular components remain aligned and functionally integrated, and any disruption of cellular homeostasis can activate the pathways that result in the destruction of the cell and all its components. The unrelatedness of the different prokaryotes and their close structural association poses interesting questions about selection. In the early stage, reciprocal altruism may have been a driver of cooperation. Later, as the cells became more functionally and structurally integrated, group selection was possible (see chapter 16). There are other hypotheses for eukaryogenesis like predator-prey interactions, but these are not included here because they are less developed.

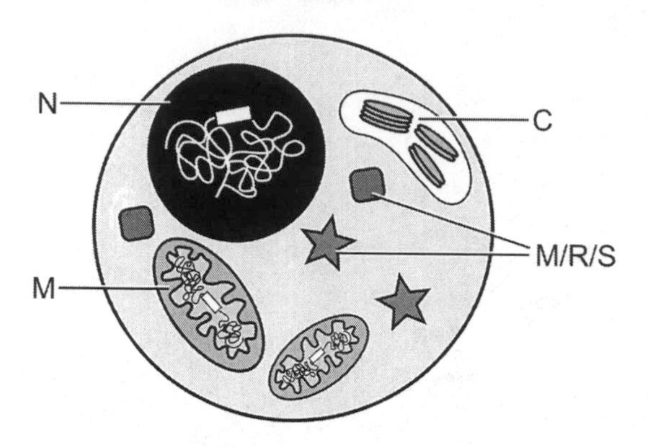

FIG. 26. The eukaryote cell. The emergence of the eukaryote cell aligned the evolutionary interests of different prokaryotes that gave rise to the nucleus (N), mitochondrion (M), and chloroplast (C). PCD was a conflict mediator between the three unrelated genomes. The newly formed eukaryote cell also contained metabolites, resources, and signaling molecules (M/R/S) that connected and functionally integrated the new organelles.

STEP 7: COOPERATIVE GROUPS OF UNICELLULAR EUKARYOTES (FIG. 27)

Just as prokaryotes form cooperative groups, so do unicellular eukaryotes. They may provide information to others about the environment (see the example of nitrogen oxide in chapter 10), communicate, and share resources. The social behaviors like cooperation and altruism in populations of eukaryote cells can again be driven by reciprocity or kin or group selection. The selection for PCD, however, would have depended on kin or group selection. Some of the bacterial PCD mechanisms that are implicated in steps 4 and 5 have been co-opted by eukaryotes (Koonin and Aravind 2002; Gao et al. 2019), while others would have evolved independently.

STEP 8: MULTICELLULAR BEHAVIOR IN UNICELLULAR EUKARYOTES (FIG. 28)

In some instances, cooperative cell groups go beyond simply sharing resources and intercellular communication. Cells can form multicellular-like structures as part of their life cycle, with one of the most investigated taxa being *Dictyostelium*. Slime molds exist mostly as free-living amoebae, but during periods of environmental stress, individuals aggregate to form

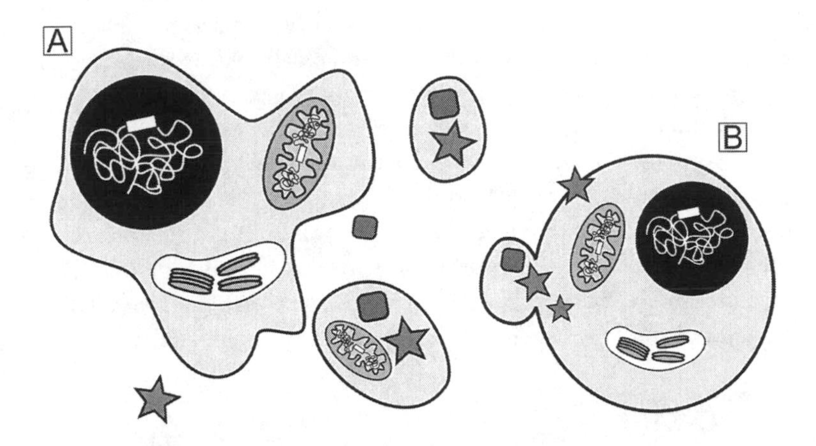

FIG. 27. Cooperative groups of unicellular eukaryotes. (A) A eukaryote cell dying by PCD releases resources or signaling molecules as well as membrane-bound apoptotic bodies that (B) can be taken up by others. The transferred materials enhance the fitness of the recipients.

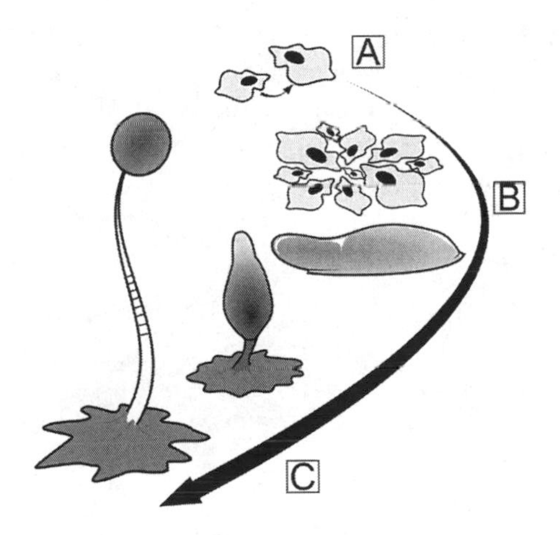

FIG. 28. Multicellular behavior in unicellular eukaryotes. In *Dictyostelium*, (A) cells may exist as free-living amoebae. PCD can benefit others as described in step 7. (B) Under environmental stress, amoebae aggregate and form a migratory slug. (C) The slug differentiates into a stalk and spore-forming fruiting body. PCD typically occurs in the stalk and is essential for the formation of the stalk and fruiting body. Without PCD the entire colony is non-viable.

a migratory slug. PCD plays a role in both unicellular and multicellular stages. In the free-living stage, healthy individuals take up the membrane-bound corpses of cells that die by PCD (Arnoult et al. 2001). This is presumed to be a way of sharing resources or genetic exchange. In the multicellular stage, PCD is essential for the completion of the organism's lifecycle (Otto et al. 2003). Without PCD, the stalks and fruiting bodies do not form, and spore dispersal is compromised. PCD is responsible for the division of labor, the formation of reproductive structures, and dispersal and is driven by kin or group selection (again, depending on the cost-benefit ratio and the degree of relatedness in the multicellular structure).

The minor steps in PCD evolution

There are many variations on the eight steps listed above but for the most part, these do not seem to involve significant evolutionary innovations, particularly in terms of the level of selection. Rather, they are modifications or co-options of preexisting processes and mechanisms. For example, in step 1, TA modules are certainly not the only examples of "selfish genes" that induce cell death. In steps 4 and 7, there are many variations on the theme of communication and resource sharing. The important evolutionary innovation in these steps, however, was the development of group-level behavior and the subsequent occupation of new niches and the evolution of adaptive pathways.

PART THREE

Origins of life and death, and their coevolution

There is nothing either good or bad, but thinking makes it so.

WILLIAM SHAKESPEARE, *Hamlet*, Act II, Scene II

15
Group selection and the origins of life and death

Evolutionary biology is the organizing framework that explains life. Whether the focus of investigation is a single taxon, interactions between taxa, or the dynamics of communities, the same evolutionary ecology processes like competition, predation, cooperation, etc., can occur at any level of inquiry. Just like any other phenomenon in the living world, the origins of both life and death have been shaped by these processes. The similarities in the sociobiological processes that occurred at the origins of life and death (see the syntheses in chapters 7 and 14) are perhaps the most intriguing because the very existence of some of the phenomena themselves have been a source of so much controversy. Of these, group selection has one of the most contentious histories in all of biology. It is ironic that such a controversial topic appears to be important for two of the most fundamental biological processes—the origins of life and death.

Before considering the role of group selection, it is important to know what is meant by the term (West, Griffin, and Gardner 2007). As discussed in chapter 13, group selection and kin selection are sometimes considered equivalent and from a purely mathematical, functional perspective this is true (Lehmann et al. 2007; Marshall 2011). But the causal structures of the two processes are fundamentally different (Birch and Okasha 2015; Kramer and Meunier 2016; Okasha 2016) and this distinction is important when considering how these two driving forces shaped the origins of life and death. Kin selection adopts a gene's eye view and the causal relationship with selection rests on the genetic structure of the group or population. Kin selection depends on understanding the genetic relatedness between individuals and Hamilton's approach of inclusive fitness is typically used to formulate the process. Group selection, on the other hand, does not have this same requirement and the theory claims that natural selection can sometimes act on whole groups of individuals. Some groups can be favored over others because of their group-level traits. In other words, there is selection of traits that are group-advantageous, and the genetic structure of the group is not the focus, even if relatedness assists with selection. A second important distinction is between group selection and reciprocal altruism. Both phenomena explain cooperation or altruism, but in reciprocity selection acts at the level of the individual, as opposed to the group (Ramsey and Brandon 2011; Birch 2017).

The role of group selection has been somewhat surprising in our appreciation of the origins of life and death. Despite the controversy surrounding group selection theory, it is claimed that group selection was important, and perhaps even essential, for both life and death to emerge.

Group selection and the origin of life

The theoretical data (chapters 2 and 6) and the synthesis for the origin of life (chapter 7) invoke group selection in one form or another. The ideas around hypercycles and quasispecies (Eigen 1971; Eigen and Schuster 1977), the "first replicators" (Michod 1983), the stochastic corrector model (Szathmáry and Demeter 1987) and many others, all make use of selection at the level of the group. The authors may formulate the origin of life process in different ways, but group selection is an important component. In the case of hypercycles, an argument can be made that reciprocal altruism is the better explanatory framework because the altruistic activity of one kind of replicator eventually comes back to itself. But I suspect that this argument will unravel when the populations are structured in such a way to exclude parasitic elements (chapter 3). It is interesting that group selection is consistent with the philosophical approaches (chapters 1 and 2) and empirical data (chapter 5), even though the experiments were not framed in these terms. In the case of the self-sustaining system of two ligases and four substrates developed by Lincoln and coworkers (Lincoln and Joyce 2009), enforced reciprocity is initially the more accurate description, but once additional ribozymes are added to the system, group selection is more appropriate. Certainly, in the interpretation of the origin of life as an ETI, group selection is invoked. It seems that, irrespective of the proposed mechanisms responsible for the first living systems to emerge, group selection played a part at some stage of life's origin. Selection acted on the properties of the group (see for example step 7 in chapter 7). The genetic structure of individuals in the group, while playing a part, was not the primary driving force. In fact, in many of the theoretical models, relatedness does not feature. The causal structure appears to be one of group selection and not kin selection.

Group selection and the origin of death

Group selection and PCD is a more convoluted issue because it deals with selection at multiple levels at the same time (genes, cells, groups of cells, the first eukaryote cell and their groups, etc.). Kin selection has clearly played a significant role in the evolution of death, and the benefits of PCD to others in the population are often directed at relatives (Durand et al. 2014; Durand, Rashidi, and Michod 2011; Fabrizio et al. 2004; Yordanova et al. 2013). In some lineages, mathematical modeling predicts that "programmed cell death can evolve in unicellular organisms due solely to kin selection" (Vostinar, Goldsby, and Ofria 2019). However, this is clearly not the case in *all* lineages or taxa. In some circumstances, relatedness is not the most significant parameter. A good example is the *E. coli* experimental system (Refardt, Bergmiller, and Kümmerli 2013) discussed in chapter 13, where the degree of relatedness was unimportant for PCD to evolve. In this group-level competition experiment, one population of bacteria drove another to extinction only if the frequency of the PCD trait in the first population was above a certain threshold, irrespective of the genetic structure of the population. Quantitative measures of relatedness were not recorded, but with relatedness being low, the frequency of PCD would also have had to be low for the $(c/b) < r$ rule to hold. That, however, was not the case.

There is a much more obvious example of group selection associated with PCD. This is the evolution of the eukaryote cell where the new kind of individual (the eukaryote cell) emerged from ancestral prokaryote cells (Blackstone 2013, 2016; Norris and Root-Bernstein 2009). There are several hypotheses for eukaryogenesis, for example, predator-prey interactions, parasitism, and endosymbiosis, but perhaps the most widely held belief is that the relationship between the different prokaryotes was one of cooperation (Blackstone 2013; Eme et al. 2018; Michod and Nedelcu 2003, 2004; Norris and Root-Bernstein 2009). The first eukaryote cell was not only physically different (in size, structure, and function), but it incorporated metabolic elements of the prokaryotes from which it emerged. In eukaryogenesis, the selection pressures driving cooperation appear to be associated with both reciprocal altruism and group selection. In the early stages, bacteria and archaea may have complemented each other metabolically, thus increasing their fitness. This is reciprocity. However, at some stage the group-level properties become important. There were emergent properties in eukaryotes that allowed them to occupy new ecological niches, with subsequent selection for these traits. But, perhaps the most important argument for group selection rests on the introduction of PCD as a conflict mediator. PCD in one of the individuals in the group rules out reciprocal altruism as a possibility because it enforces selection of the group-level traits (Blackstone 2013, 2016; Blackstone and Green 1999).

At some stage of eukaryogenesis, once reciprocity dissipated as a major driving force (due to group traits and loss of individuality in prokaryotes), it is argued that group selection rather than kin selection is the explanatory framework. This is because the cooperating prokaryotes were from divergent taxa. Selection was independent of relatedness and depended on group-level properties. In fact, the absence of any relatedness between prokaryote genomes meant that conflict was inevitable, and kin selection would have worked against the evolution of the eukaryote cell. The exchange of genetic material between organelles (nucleus, mitochondria, and plastids) ameliorated the situation, but the potential for conflict remained and stills plays out in extant eukaryote cells (Burt and Trivers 2006). If cooperation was the key process in eukaryogenesis, then reciprocity and group selection were the most likely drivers of eukaryote evolution and the primary conflict mediator was, and still is, PCD (Blackstone 2013, 2016; Blackstone and Green 1999; Kaczanowski, Sajid, and Reece 2011; Michod and Nedelcu 2004; Norris and Root-Bernstein 2009).

Group selection of aggregate and emergent properties at the origins of life and death

Group traits that are the target of natural selection may be aggregates (of the individuals making up the group) or emergent (properties associated with the group and irreducible to individuals within the group) (Thompson 2000). An example of group selection acting on the aggregate is the case where the frequency of PCD in the group crosses a threshold such that the group outcompetes another group in which PCD is less frequent or absent. The data from the *E. coli* and *L. major* experiments discussed earlier provided the empirical support.

The emergent property of the group can also be selected for, and there are several examples of this at the origins of life and death. As discussed in chapter 6, the origin of life depended on the selection of groups of molecules and groups of LRUs (lower-level replicating units; additional notes 15.1). In many instances the group-level trait selected for was emergent and Eigen and Schuster's (Eigen and Schuster 1977) quasispecies, Michod's (Michod 1983) first replicators, and Szathmáry and Demeter's (Szathmáry and Demeter 1987) replicating protocells, in one way or another, can all be interpreted in this way. But perhaps the best (or most obvious) example of a group where the emergent property is selected for, is the abstrac-

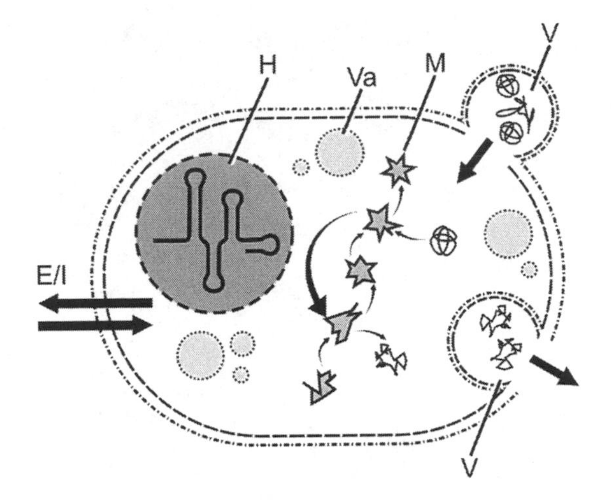

FIG. 29. The protocell. The image is reproduced from fig. 16 in chapter 7. The protocells emerged from HRUs and contained hereditary (H) components (ribozymes, RNAs, simple amino acids, and cofactors) and a primitive metabolism (M). There is efflux and influx of simple compounds (water, salts, etc.). Vesicles (V) may enter or leave the protocell, and vacuoles (Va) are intracellular stores of metabolites, salts, water, etc. There is an efflux and influx (E/I) of nutrients, salts, water, and lipids between the protocell and the environment. The cell may still contain ribozyme elements that have the potential for independent replication, but the emergent properties of the cell overcome selection at the level of individual ribozymes or their networks.

tion of a protocell. In the earliest cell-like individuals there were traits that emerged from the functional activities of individuals within the protocell (or its precursor) (fig. 29). There may still have been components in the protocell that functioned as individuals in their own right, but their collective activities led to emergent traits. These may have included simple metabolic pathways, lipid biosynthesis, or the maintenance of hereditary material. An analogy can be found in Gánti's chemoton, where the cell is maintained by the activities of the hereditary material and its products. It should be remembered, however, that in the chemoton the hereditary material is not conceptualized as discrete individuals in a group, whereas the LRUs that cooperated to form an HRU (higher-level replicating unit) are individuals in their own right.

Group selection of an aggregate or emergent property at the origin of death can be found in cooperation in bacteria, between bacteria and archaea, and in multicellular behavior in amoebae. Examples of these are bacterial biofilms (fig. 25), eukaryogenesis (fig. 26), and collective behavior in groups of eukaryote cells (fig. 28).

16

Life and death coevolution, and the emergence of complexity

Despite the intuitive notion that life and death are antonyms, and of course in many ways they are, death has also augmented life in more fundamental ways than previously imagined. An important distinction needs to be made at the outset. In a purely superficial sense, death clearly has the side effect of facilitating life. The decomposition of a corpse provides nutrients that allow other organisms to thrive. Clearly, an individual does not die *so that* it can nourish other forms of life, and the relationship between life and death is one of circumstance. These instances are distinct from other situations in which life and death have coevolved and are coadapted, where life and death not only depend on each other but have impacted each other's evolution. In these cases, death has played an unexpected role in the evolution of more complex life (the issue of complexity is dealt with in part 1; see also Adami 2002; Cilliers 1998). Life and death are not only inter-dependent, they exhibit coevolutionary features.

Meanings of coevolution

The more traditional interpretation of coevolution refers to the reciprocal evolutionary relationship between two or more species. The idea can be traced back to Darwin, when he described the relationship between flowers and their pollinators (Darwin 1862). Then there was the seminal work by Ehrlich and Raven, who cemented the concept in their work on butterflies and plants (Ehrlich and Raven 1964). But in the modern era, coevolution has taken on a much broader meaning. The concept no longer just refers to the ways in which two species may coadapt. There is co-

I am very grateful to Rick Michod for his reading, corrections, and comments on this chapter. At the same time any errors are mine alone.

evolution at the level of molecules, organelles, cells, and even at a systems biology level (Carmona, Fitzpatrick, and Johnson 2015; Nuismer 2017). Coevolution rests on the idea that two biological entities, whether they are organisms, traits, molecules, etc., interact in such a way that they impact each other's evolution. With respect to life and death, the question is whether the PCD trait has been a causal influence in the evolution of different kinds of life forms (for example, unicellular, multicellular, or social insects) and whether the reverse is also true. Has the evolution of life been a causal influence on the evolution of the PCD trait?

Life-death coevolution in microbial systems

In the microbial world, a simple example of PCD coevolution in a community is found in chlorophyte-archaebacterium environments (Orellana et al. 2013). *Halobacterium salinarum*, for example, induces PCD in the alga *Dunaliella salina* by secreting molecular lysins and the archaebacterium lives off the algal lysate. *D. salina* in turn regulates PCD in itself, releasing nutrients into the environment that are taken up by the archaebacterium and other *D. salina* cells. The chlorophyte and archaebacterium have coevolved through a molecular interaction that leads to the induction or inhibition of PCD in the alga. The interactions in the *Dunaliella-Halobacterium* community present in the Great Salt Lake have been dissected in detail (Orellana et al. 2013), but there are other similar examples where phytoplankton and prokaryotes have evolved interactions via PCD mechanisms (Bidle 2015, 2016; Franklin, Brussaard, and Berges 2006). The physiological interactions and metabolic connections (either via PCD or other mechanisms) between microbes are broadly referred to as the microbial loop (Fenchel 2008). When the interactions are mediated by PCD, death is the mechanism responsible for the coadapted connection between two or more taxa (additional notes 16.1). The chlorophyte response appears to be coadapted at two levels of selection. At the organism level, the chlorophyte has adapted to the PCD-inducing molecules released by the archaeon. At the kin / group level the induction of PCD supports the nutritional requirements of relatives. The experiments by Orellana and colleagues (2013) revealed that the responses have fitness effects (they are adaptive), but whether all the fitness-related benefits have an evolutionary history of selection (in other words whether they are an adaptation and hence coevolved) was not examined.

Life-death coevolution in microbes extends to host-virus interactions, where a coevolutionary arms race is mediated through PCD mechanisms. The host and virus battle for control over cell fate via a molecular-level arms race (Schwartz and Lindell 2017; Gao et al. 2019; Vardi et al. 2012; Vardi et al. 2009). Such interactions are well documented in viruses and their phytoplankton hosts (Vardi et al. 2012; Vardi et al. 2009). In some instances, such as the *Prochlorococcus*-podovirus example, the interaction has been dissected in great detail at the molecular level (Schwartz and Lindell 2017).

Life-or-death choice in prokaryotes is coupled to individual immunity (Koonin and Zhang 2017). Ironically, the evolution of PCD facilitates the survival of others in the group at the expense of the individual (Refardt, Bergmiller, and Kümmerli 2013). In other words, life and death struggles at the individual and group levels are mediated by PCD. These and other examples reveal that microbial life in the community is causally impacted by cell death. The converse is also true. The evolution of life affects the mechanisms underpinning death. But the most profound ways in which living systems and cell death mechanisms have coevolved can be seen in the evolution of more complex life (Durand, Barreto Filho, and Michod 2019).

PCD and the evolution of the eukaryote cell

PCD in eukaryote cells was introduced predominantly via the true bacteria, which are understood to be the ancestral forms of mitochondria. Koonin and Aravind discovered this "bacterial connection" by identifying diverse homologs of PCD proteins in bacteria, but not in archaea (Koonin and Aravind 2002). There may be other mechanisms of PCD in archaea, including caspase-related pathways (Bidle et al. 2010; Seth-Pasricha, Bidle, and Bidle 2013), but there is a strong bacterial connection with the PCD mechanisms of eukaryotes. The phylogenetic and genomic data are supported by laboratory studies, which revealed that the mitochondria play a central role in apoptosis (the commonest phenotype of PCD) in eukaryote cells (for example, Kroemer and Reed 2000). The circumstances that drove the interactions between different prokaryote taxa are not entirely clear. As discussed earlier (chapter 15), one of the hypotheses for eukaryogenesis is that the cooperative and eventually obligate, symbiotic relationship with bacteria conferred the amitochondriate host cell with new metabolic capabilities, in particular oxidative respiration (Blackstone 2013; Frade and Michaelidis 1997; van der Giezen 2011). The later events that led to a similar relationship with cyanobacteria provided the advantage of photosynthetic energy. Chloroplasts, the organelle descendants of cyanobacteria, are also involved in PCD and many of the processes described in mitochondria are also applicable to chloroplasts. The genetic differences between the two taxa that gave rise to the amitochondriate cell and the mitochondria would have inevitably led to conflict (Blackstone 2016; Michod and Nedelcu 2004), and Blackstone and Green suggest that "a mechanism of apoptosis in metazoans may thus be a vestige of evolutionary conflicts within the eukaryotic cell" (Blackstone and Green 1999). A similar argument is made by Kaczanowski (Kaczanowski 2016), and the ancestral state reconstruction of the eukaryote cell by Klim and colleagues found "an ancient evolutionary arms race between protomitochondria and host cells, leading to the establishment of the currently existing apoptotic pathways" (Klim et al. 2018). This latter finding was supported empirically in their yeast model system.

The molecular coevolution between the ancestor of mitochondria and the host cell eventually resulted in the integration of cell death pathways. However, the potential for conflict between the genomes of the different cellular organelles remains (Burt and Trivers 2006), and as Michod and Nedelcu have indicated (2004), PCD can mediate this conflict. In the eukaryote cell, PCD is a way of ensuring that the evolutionary interests of all the organellar genomes (three, in the case of photosynthetic eukaryotes) remain aligned. The stable evolution of the eukaryote cell depended on this coevolution of life and death. In an unexpected way, the evolution of cell death in bacteria was essential for one of the great evolutionary transitions in individuality, the emergence of the eukaryote cell.

PCD and the evolution of multicellularity

PCD is connected to another evolutionary transition in individuality and complexity, the evolution of multicellularity, in a surprising number of ways (table 3). Some investigators have highlighted its role in the evolution of multicellularity, describing "the evolution of cell death programs as prerequisites of multicellularity" (Huettenbrenner et al. 2003). The investigations of PCD in unicellular organisms were aimed at uncovering the reasons for this trait in the microbial world, but they also provided the first clues that PCD was important for the emergence of multicellular life. The mathematical models developed by Iranzo et al. revealed that "the joint evolution of cell aggregation and PCD is the optimal evolutionary strategy" in response to a virus-host arms race (Iranzo et al. 2014). Many of the experiments in unicellular organisms examined the potential kin / group-level effects of PCD and, in a range of organisms, from phytoplankton to yeast, kinetoplastids, apicomplexa, and amoebae, PCD enhances the fitness of others. Functionally, this is consistent with the early steps to multicellularity where kin / group-level fitness drives the altruistic behavior of individuals in the social group (Buss 1987; Michod 1999;

TABLE 3. PCD and the evolution of multicellularity

PCD-dependent mechanism required for multicellularity	**Examples in model organisms**
Resource sharing, communication, and group-level survival	*Saccharomyces, Chlamydomonas, Leishmania, Escherichia, Peridinium, Dictyostelium*
Dispersal, and spore / propagule formation	*Dictyostelium*, group-level propagation in the "snowflake yeast"
Division of labor	*Dictyostelium*, "snowflake yeast"
Transfer of fitness	*Saccharomyces, Chlamydomonas, Dictyostelium, Leishmania, Escherichia, Dunaliella*
Conflict mediation	Eukaryote cells in unicellular or multicellular organisms

Note. PCD, programmed cell death.

Okasha 2006). As Zuppini et al. say, the adaptive explanation for PCD in single-celled organisms is "based on the concept that unicellular life could be able to organize itself into cooperating groups" (Zuppini, Andreoli, and Baldan 2007).

In some model organisms that exhibit both unicellular and multicellular stages, PCD is key to the multicellular stage. In *Dictyostelium* spp., which have alternating unicellular-multicellular lifecycles, PCD is required for the development of the multicellular stage. In *D. discoideum*, mutants that do not have a PCD mechanism are unable to produce stalks or spores (Otto et al. 2003). The single-celled amoeba stage appears unaffected, but the abnormal stalk and spore formation compromises reproduction and dispersal. These findings demonstrated that PCD was a mechanism for the division of labor in multicellularity (Cornillon et al. 1994). Other taxa exhibit a similar reliance on PCD. In bacteria like *Myxococcus* (Kaiser 1986) and *Streptomyces* (Filippova and Vinogradova 2017) differentiation and development in the multicellular stage depend on PCD. Propagation and dispersal depend on PCD in *Trichodesmium*, a filamentous diazotrophic cyanobacterium. When the environment is unfavorable, some cells in the filaments of cyanobacteria differentiate into dying cells, leading to a fragmentation of the colony (Berman-Frank et al. 2004). The dispersal of these propagules, called hormogonia, plays a role in colonizing new environments and bloom development when environmental conditions improve. While the role of death at the unicellular-multicellular transition has been discovered in naturally occurring organisms, a similar scenario can be observed experimentally. The importance of PCD for propagation was proposed in the experimental evolution of groups of yeast cells—the "snowflake yeast" model system (Ratcliff et al. 2012). Just as in the colonial phytoplankton *Trichodesmium*, new colonies broke from the parent multicellular yeast at sites of PCD cells. This allowed the propagation of the snowflake form. The physical properties of the multicellular structures were also responsible for the "fracture[s] due to crowding-induced mechanical stress" (Jacobeen et al. 2018). The authors argued that both physical fracturing and PCD resulted in colony propagation and that "apoptosis is a trait that coevolves with large cluster size" (Pentz, Taylor, and Ratcliff 2016). Dissecting the causal chain of events will require additional information, but either way, cell death can facilitate propagation of the multicellular stage in this experimental system.

PCD is linked to the origin of specialized cell types in a range of multicellular lineages. It is "an essential process of cereal seed development and germination" (Dominguez and Cejudo 2014) and the origin of new cell types in metazoan development has been attributed to cell stress pathways and PCD. It is argued that environmental stress can lead to the differentiation of specialized cells and that this is achieved via PCD pathways—a phenomenon called stress-induced evolutionary innovation (Wagner, Erkenbrack, and Love 2019). For example, the decidual cells that reside in the uterus of eutherian mammals have evolved by way of PCD pathways (Erkenbrack et al. 2018), and skin thickening in cetaceans is a stress-induced PCD mechanism (Eckhart, Ehrlich, and Tschachler 2019). In these cases, oxidative stress and PCD are required for the differentiation of specialized cell types that are not only important, but essential for the viability or reproduction of the multicellular organism. The terminology in these examples can lead to confusion because it can be asked: How can a dead cell differentiate into a living specialized cell if it has died? This is one of the idiosyncrasies of the terminology of death used in the literature. The issue is covered in chapter 8, but death is a "spectrum" with a "range of phenotypes" (Thomas et al. 2003) and there are "degrees of dying" (Durand and Ramsey 2019).

In addition to group-level viability and division of labor, PCD is a mechanism for "the transfer of fitness from individuals to the group" (Durand, Barreto Filho, and Michod 2019). This is a concrete example of the otherwise abstract concept of fitness transfer (Michod 2005). The transfer-of-fitness view in ETIs is sometimes criticized by some philosophers of biology as having purely epistemic value (Bourrat 2015), but it is helpful for interpreting some important observations. For example, in *Dictyostelium*, membrane-bound packages of cellular material (apoptotic bodies) are the result of PCD and have been observed being engulfed by healthy neighbors (fig. 2D in Arnoult et al. 2001). Tatischeff et al. observed similar apoptotic bodies in *D. discoideum*, finding them to contain DNA (Tatischeff et al. 1998). The equivalent structures of apoptotic bodies have been observed in *Saccharomyces* (Madeo, Frohlich, and Frohlich 1997) and *Chlamydomonas* (Durand, Sym, and Michod 2016; Moharikar et al. 2006), although their potential uptake by other cells was not investigated. The materials released by PCD that benefit others are not confined to membrane-bound structures (Durand, Rashidi, and Michod 2011) and in *Dunaliella*, they take the form of energy-rich resources that move from dying cells to healthy ones (Orellana et al. 2013). This "access to resources" (Hochberg, Rankin and Taborsky 2008) is an essential component in enhancing group-level effects. The cycling of dissolved organic materials in marine microbial communities is widely documented (Bidle 2015, 2016; Franklin, Brussaard, and Berges 2006) and microbial interactions in the ocean surface (the "sea skin") are sometimes used as a model system to investigate the key features associated with multicellularity (Abada and Segev 2018).

PCD is one of the conflict mediators in multicellularity evolution (Michod 2003; Michod and Nedelcu 2004). Altruistic death is, indeed, the ultimate way of removing conflict between individual cells in the group. In multicellular organisms the cells are usually clones (or near clones) of each other. Early in the evolution of colonial living, however, the genetic relationships between individuals in the group depended on how the groups arose. Unicellular organisms are known to exhibit predator-avoidant phenotypes and behaviors (for example, Lürling 2003), and Stanley suggested that predation was a major driver of multicellularity in general (Stanley 1973). Several unicellular green algal taxa have subsequently been used to investigate this hypothesis with different emphases, first *Scenedesmus* (Lürling and Van Donk 1996), then *Chlorella* (Boraas, Seale, and Boxhorn 1998) and *Chlamydomonas* (Becks et al. 2010; Herron et al. 2019). In some instances at least (Boraas, Seale, and Boxhorn 1998; Becks et al. 2010; Herron et al. 2019), the colonial form is a stable, heritable phenotype (but see additional notes 16.2).

Some researchers assume clonality as a starting point in the evolution of multicellularity and, of course, that may be the case in many instances. The assumption is presumably because kin selection is such a key driver of coloniality, and clonality is usually a feature of extant obligate, "true" multicellularity forms (additional notes 16.3). It is easy to see how kin selection played a role if cells stayed together after division. But what about the cases where cells come together (the phrases "stay together" and "come together" are used in the literature to describe the two ways in which cell groups arise)?

In *Dictyostelium*, "multicellular" slugs are cell aggregates that form by individuals coming together. In wild isolates and laboratory experiments the multicellular stage can comprise genetically heterogeneous cells (Fortunato et al. 2003; Sathe et al. 2010; Strassmann and Queller 2010), which is a potential source of genetic conflict. High relatedness and recognition systems (Gilbert et al. 2007; Gruenheit et al. 2017; Kundert and Shaulsky 2019), of course, do facilitate cooperation, but in genetically diverse colonies conflicts will arise, depending on the degrees of relatedness. In *Chlamydomonas* predator-induced aggregates may comprise different species (Sathe and Durand 2016) (fig. 30) and even different genera (Kapsetaki, Fisher, and West 2016). The multicellular groups are formed by individuals both coming together and staying together after division (Kapsetaki, Tep, and West 2017). Putting these findings together, the following emerges: (i) predation is a driver for the formation of groups of cells (the argument can be made that predation was the dominant selective pressure driving multicellularity; Stanley 1973; Boraas, Seale, and Boxhorn 1998; Lürling and Van Donk 1996; Solari, Galzenati, and Kessler 2015); (ii) under predation pressure, groups can comprise genetically unrelated individuals; and (iii) if, in some cases, heterogeneous groups like these gave rise to multicellular organisms, how was the problem of genetic conflict overcome? How were non-relatives excluded from the group?

Some researchers may reject the idea that multicellularity could have evolved from groups of different species or strains, largely because of the problem of genetic conflict. However, the possibility that true multicellularity could have evolved from genetically heterogenous groups should not be dismissed out of hand. In the green algal model systems, different species coexist and Sathe, Kapsetaki, and others demonstrated that predator-induced unicellular aggregates can be chimeric if more than one taxon is present in the community. Whether such aggregates can eventually form stable colonies is another question, but under extreme artificial selection pressures that focus on one trait at the expense of others that can impact viability (for example, Boyd et al. 2018), this cannot be ignored. There is also no reason a priori to assume that clonal groups are the starting point from which multicellular life emerged. The causal process must be distinguished from the products it generates. Clonality in multicellular life is evidence of kin selection but it is not a precondition. Certainly, it seems easier for clonal groups to evolve multicellularity, but to reject the possibility that multicellular life evolved from chimeric colonies because of the is-

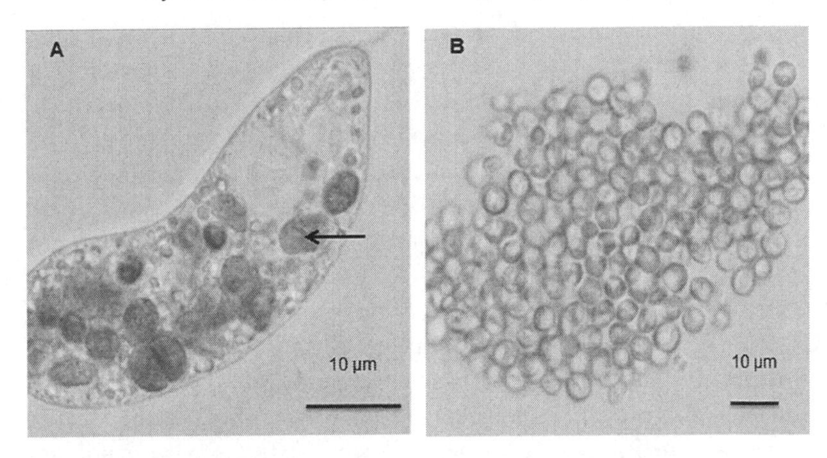

FIG. 30. Predator-induced aggregation. (A) *Peranema trichophorum* (predator) with ingested *Chlamydomonas reinhardtii* algal cells. (B) When they sense the presence of predators, algae avoid being ingested by aggregating to form groups. (Image courtesy S. Sathe.)

sue of genetic conflict is not correct. To borrow Williams' phrase, to make that assumption would be "putting the cart before the horse" (Williams 1992). What needs to be explained is how the problem of genetic conflict may have been overcome *if* multicellularity evolved from groups that arose from aggregation, that is, cells "coming together." How did groups comprising related individuals arise from chimeric groups?

There may be several solutions, one of which is PCD, to the problem of genetic conflict in mixed groups. The mathematical models of Bouderbala et al. indicate that, for a group of phytoplankton cells to remain viable under realistic environmental conditions, there must be some form of cell death occurring in the group (Bouderbala et al. 2018, 2019). An earlier mathematical model by Iranzo et al. predicted that cellular aggregation and the evolution of PCD is the optimal strategy in response to viral invasion of a population by unicellular organisms (Iranzo et al. 2014). Theoretical studies like these did not include predation as a driving force but they do point to the importance of PCD in the early stages of group formation. In most of the experimental systems investigating predator-induced aggregates mortality was not reported. For example, the early experiments by Boraas (Boraas, Seale, and Boxhorn 1998) and Lürling (Lürling and Van Donk 1996) and the growth rate analyses of groups of *Phaeocystis* cells (Veldhuis, Brussaard, and Noordeloos 2005) did not report PCD. Perhaps this was not explicitly examined or in the ideal, nutrient-rich study systems PCD did not occur. Of course, there is also the possibility that PCD would not have occurred regardless of the culture method, but in the Bouderbala (2018) and Iranzo (2014) theoretical simulations and in Sathe's experiments with *Chlamydomonas* and the predator *Peranema* (Sathe and Durand 2016), cell death as well as cell growth and division were observed. Rather than death being a problem, however, it is also an opportunity around the issue of genetic conflict in aggregates. PCD benefits clonal relatives but is harmful or neutral to other species and strains (Durand et al. 2014) depending on their relatedness. PCD, therefore, may lead to the genetic homogenization of the group, allowing kin selection to play a more dominant role. PCD in one of the genotypes will promote the growth of its relatives and inhibit the growth of others. This will allow chimeric groups to evolve into ones comprising genetic relatives, from which obligate forms of multicellularity arose. It seems to me that if predation was the dominant selective pressure driving colonial lifestyles, then some mechanism for dealing with the inevitable genetic conflict en route to multicellularity would have been required. The evolution of PCD is a potential solution (Iranzo et al. 2014; Bouderbala et al. 2018).

In metazoa (multicellular animals), the coevolution between life and death at the molecular and cellular level is widely documented. One of the clearest examples is the case of the *TP53* gene, which is a master regulator and inducer of PCD. In mice (and other model animals) the gene can be knocked out, which leads to premature organismal death. In the case of the TP53 mechanisms, PCD is required for developmental processes in the growing embryo and for inducing death in cancerous cells. The mice die because cells in the organisms do not express the PCD trait. From this example alone, it is clear that cell death and cell life have coadapted in animals. Furthermore, homologous genes for PCD have been identified across all the metazoan lineages as well as other multicellular lineages (Aravind, Dixit, and Koonin 1999) and a strong argument can be made that PCD was essential for the origin of multicellularity itself (Huettenbrenner et al. 2003). PCD aligns the evolutionary interests of individual cells, and Blackstone (Blackstone 2013, 2016; Blackstone and Green 1999), Michod (Michod 2003), Michod and Nedelcu (Michod and Nedelcu 2004), Durand et al. (Durand, Barreto Filho, and Michod 2019), and others list PCD as an example of a conflict mediator in their treatments of multicellularity and evolutionary transitions in individuality.

PCD in social insects

In the discussion above, PCD is an adaptation to a multicellular lifestyle, but there are also cases where PCD has evolved because it diminishes the fitness of the entire multicellular organism. This is the result of another cross-level adaption and proved essential for the ETI from asociality to eusociality. In some insect colonies, PCD is again a mechanism for decreasing individual fitness and boosting group fitness. The mechanistic basis for this has been dissected, quite literally, in eusocial honeybees by observing the fate of individual cells in workers. The queen secretes pheromones that induce PCD in the ovaries of worker bees. Worker oocytes are aborted, and the individual bees are effectively sterile in service of the colony. Without PCD in workers they are not rendered sterile and the social complexity of the community is disrupted. In a sense, this control of lower-level individuals is akin to somatic cells in multicellular organisms (Ronai, Oldroyd, and Vergoz 2016). Just as somatic cells in a multicellular individual give up their reproductive potential, thus enhancing organism fitness, so do some multicellular individuals give up the potential for reproduction in favor of the eusocial group. Again, PCD plays a role in a major ETI and is a necessary mechanism that allows life in eusocial colonies to thrive.

POD (programmed organismal death) in insects and arachnids

In the 1860s, Wallace suggested that aging and death might be evolved traits. He claimed that individuals are programmed to die so that they do not compete with their offspring (Wallace 1889). His idea had some early support, but by the 1920s, this thinking was derided as a "perverse extension of the theory of natural selection" (Pearl 1922). The evolutionary explanations for senescence shifted to theories like mutation accumulation, antagonistic pleiotropy, and disposable soma (see Travis 2004 and references therein). But the wholesale rejection of Wallace's idea may have been premature and in the 1980s and 1990s there was a mini-revival. McAllister and colleagues have shown that pea aphids exhibit adaptive behavioral suicide when infected by a parasitoid wasp (McAllister and Roitberg 1987), and Andrade demonstrated sexual selection for male sacrifice in the Australian redback spider (Andrade 1996). Toyama's experiments with the foliage spider revealed that mothers stimulate their spiderlings to cannibalize her before dispersal (Toyama 2001). Spiderlings that do so are fitter than those that do not. In a more recent study, "a new and astonishing case of adaptive self-sacrifice in a polyembryonic parasitic wasp" was discovered (Otsuki et al. 2019). In this instance mass killing was adaptive because the killing increased indirect fitness by promoting the reproduction of their clones. These examples showed that POD can, at the very least, be adaptive. Using van Valen's distinction between adaptation and adaptive (see chapter 8), it is not clear whether all these behaviors have themselves been selected for and researchers argue about how the data should be interpreted. Humphreys and Ruxton provide a helpful discussion of some of the examples above (Humphreys and Ruxton 2019). Sexual selection in the redback spider and self-sacrifice of male larvae in the polyembryonic wasp are perhaps the best of examples of POD with an evolutionary history that suggests selection and, in these cases at least, the behaviors of the adults and offspring appear to have coevolved.

Life-death coevolution

In curious and unexpected ways, death (PCD and in a few instances POD) impacted the evolutionary trajectory of life. The opposite is also true because as new forms of life emerged, these impacted the ways in which cells (or organisms) die. In the broadest sense, this is demonstrated by the recycling of nutrients in microbial communities. The community structure and evolutionary ecology of microbes depend on the way in which cells in the community die, and their death in turn affects the life-history evolution of others.

A much more intimate relationship between life and death is evident in the evolution of cell groups, the eukaryote cell, and multicellularity. The viability of any eukaryote cell rests on a delicate balance between the biochemical activators and inhibitors (pro- and anti-apoptotic factors) of PCD (Kaczanowski 2016; Blackstone and Green 1999). These antagonistic pathways coevolved following the evolution of the first eukaryote cell, as a way of mediating conflict between the prokaryotes that cooperated to form a more complex cell (Michod and Nedelcu 2004; Radzvilavicius and Blackstone 2018; Blackstone 2013, 2016; Blackstone and Green 1999). The evolution of a eukaryote cell and its program for death impacted the trajectory of all subsequent non-prokaryote life forms. PCD is essential for multicellular life and there are no obligate multicellular forms that do not exhibit one or other of its morphotypes. Many of the key processes required for the evolution of multicellular life (table 3) are present only because of cell death and dying and living cells coevolved to maintain the integrity of the organism. The transition to eusociality in insects depends on reproductive altruism (where PCD is the mechanism) in workers. In some non-social insects and arachnids, POD seems to be part of the life-history strategy, as observed by the suicidal behavior of aphids, wasps, and spiders and possibly other taxa that have not yet been investigated from this perspective.

Van Hautegem and colleagues, in their analysis of PCD in plants, claimed that "only in death, life" (Van Hautegem et al. 2015) but this phrase is more far-reaching than originally thought. The role of death in the living world extends beyond our intuitive understanding that it is simply the opposite of life. Death is of intrinsic value to much of the complexity that is observed in the living world. It is the way we think about death that affects our understanding and appreciation of life. Hence, the caption at the beginning of part 3. In Hamlet's discourse with Rosencrantz, he says, "There is nothing either good or bad but thinking makes it so" (Shakespeare 1603).

Postface

It is not claimed that the syntheses at the ends of parts 1 and 2 (chapters 7 and 14, respectively) are a complete recapitulation of how life and death originated. For the origin-of-life synthesis, it is unlikely we will ever know how life began, but based on our current knowledge, and an integration of the data from different fields, the synthesis presented is a parsimonious explanation for how it *could* have occurred. It is surely an over-simplified version. The reality is that there were likely many other types of molecules involved that acted as cofactors and activators of the processes described. At the same time, it may be that some of the difficulties discussed in the synthesis are overstated. It is likely there were many co-occurring molecules that have not been included. It is not known which molecular components existed at the very beginning and they may have either facilitated or impeded some of the steps. Future work may uncover more of the early biomolecules, and artificial selection experiments will further elucidate the evolutionary processes. See the many references to key works in the reference list.

The origin-of-death synthesis focuses on the adaptive processes. The general framework is realistic, but at the same time, I do not expect this will be the final word or that this will settle all the debates. But, hopefully, it will assist in organizing our knowledge in this subject. The future of PCD research, from an evolution and ecology angle, is poised to enter a new and exciting phase. It is primed to embrace additional elements, which are included in what has become known as the extended evolutionary synthesis (Laland et al. 2015). These include evolvability, phenotypic plasticity, niche construction, epigenetics, and developmental bias, which are so far unexplored in unicellular PCD. As an example of how epigenetics may impact PCD evolution, see the discussion in chapter 9. Evolvability is also related to death; see Mitteldorf and Martins' works (Mitteldorf and Martins 2014) for examples, but there are many others. In addition, the levels-of-selection debate in PCD will likely enter a new phase, where holobionts are considered a bona fide unit of selection (Lloyd and Wade 2019; Roughgarden et al. 2017) and that the PCD trait can be maintained when the holobiont (comprising phytoplankton and their cooperating bacteria or archaea, for example, Stat et al. 2012) is an inherited condition as opposed to an acquired one (Roughgarden 2020). For an alternative view arguing against holobionts as units of selection see Skillings 2016.

Our appreciation of PCD is not only of academic interest. As Bidle, Vardi, Falkowski, Segovia, Franklin, Brussaard, Berges, Kaplan, Orellana, Jimenez, and many others (my apologies to the many researchers in this field that I have not mentioned) have argued, the role of PCD in marine phytoplankton, which are major primary producers and responsible for almost half of global net primary production (Field et al. 1998), is only beginning to be uncovered. The abundance and community structures of phytoplankton directly impact higher trophic levels and key biogeochemical cycles, and how we deal with the impacts of oceanic changes due to global warming will determine the health of the planet as well as our own. Algal blooms, for example, appear to be increasing in frequency (Van Dolah 2000) and can have devastating ecological consequences (for example, Ndhlovu et al. 2017). Except for the finding that PCD does play a role in phytoplankton blooms (for example, Vardi et al. 2007), its broader impact in marine microbial ecology is largely unknown. The relationship between life and death evolution may be of particular interest if (or when) extraterrestrial life is uncovered. Our first encounter with life outside of Earth's biosphere may very well be of a microbial nature. Since resources and the physico-chemical environments constrain living systems it will be interesting to see whether there is an active role for death in maintaining life.

Part 3 reveals the extent to which the evolution of life and death have impacted each other. The emergence of more complex forms of life like multicellularity depended on the evolution of PCD. At the same time, the different traits associated with PCD emerged depending upon the needs of the multicellular organism. The realization that whole organism death in multicellular life (POD or programmed organismal death) can, in fact, be programmed was unexpected. The behavioral programs that lead to organismal death are coadapted with the behavior of others who benefit from it. Aging in multicellular organisms appears to be the result, largely, of processes like antagonistic pleiotropy and other trade-offs, mutation accumulation, etc., although in some instances it is argued that aging itself can be adaptive (Singer 2016; Mitteldorf 2016).

Additional notes

The additional notes are intended to make the reader aware of some of the other notable researchers, publications, and counterarguments that are not mentioned in the main text. Some of the relevant issues are also expanded upon in a less formal way.

PART 1

Chapter 1

1.1 Individuality is a concept worthy of much more discussion. The term *individual* is used very loosely throughout this book. It is not explored in any great detail, although it is discussed further in later chapters. The references provide further reading.

1.2 Not all evolution occurs by natural selection, although that is, of course, a major force. It is useful to bear in mind that other non-adaptive processes like, for example, chance and probability (Denny and Gaines 2002; Koonin 2011; Ramsey and Pence 2016) and genetic drift (Lynch 2007) also play a role.

1.3 I am not certain who should be credited as the first person to say that "life is an emergent property." Variations of the phrase can be found in a range of academic and non-academic writings. For more of emergence see Vesterby 2011.

1.4 The term RNA world was coined by Walter Gilbert, but the ideas concerning RNA as the primordial molecule of life were developed much earlier. A vast number of important works have emerged since. For further reading on the RNA world, the references cited in this book and the additional references therein are recommended.

Chapter 2

2.1 In the non-scientific world, and very occasionally even among academic researchers, students may come across the argument that crystals have "living" properties. It is worth mentioning this to arm unsuspecting students with some of the counterpoints. Indeed, salt crystals can grow and reproduce in the sense that the basic crystalline chemical structure is repeated. However, crystals for the most part have fairly rigid structures, with covalent bonds linking the chemical elements. There is no meaningful evolution by natural selection and no life scientist, or any reasoning individual for that matter, would seriously claim that crystals are alive. The philosophical arguments in the fields of complexity and functionalism easily refute the notion of living crystals.

2.2 There may be preferences regarding the use of terminology like *law, principle, theory, theorem,* or *rule.* It does not make any difference at this stage, and for personal preference I have elected to use *principle,* but one could just as easily use *law,* to capture the fundamentals that differentiate biology from physics and chemistry.

2.3 Alexander Oparin was one of the first biochemists in the twentieth century to write specifically about the differences between biota and abiota (Oparin 1953). He often argued that there was no fundamental difference between living and non-living material but discussed this in the context of the chemical matter that composes life and non-life. Although he did not speak explicitly about emergence in the same way as it is referred to in this chapter, he often referred to some of the interactions between the chemical components as being unique to living systems. He also referred to phenomena like natural selection, adaptation, and other concepts associated with evolution as being unique to life.

2.4 For the sake of making the reader aware of alternate views, there are many thinkers who are critical of the concept of emergence and the arguments invoking emergent properties (see for example Delehanty 2005). They will argue that the reductionist approach is sufficient.

2.5 The basic concepts in evolutionary biology warrant further discussion and the reader is referred to the references listed here for a more detailed treatment. An example of one of the concepts is that of the "unit" of life. The term *unit* is used intentionally here. Loosely speaking, units may be individuals or organisms, but as with so many concepts in evolutionary biology, the word is a loaded one and comes with a long history. The words *individual* and *organism* are also loaded, but it is impractical to delve more deeply into the concepts and terms in evolutionary biology in this book. The references are suggested for further reading.

There is another deeper philosophical issue, namely causality in evolution, that I am ignoring. It cannot be dealt with in any meaningful way here; however, it is at the heart of the philosophy of evolutionary theory and it would be remiss not to make the reader aware. For the sake of simplicity and argument, I am taking the position that natural selection is a causal process and not the product of an evolutionary process. For more on this key debate see (Ramsey 2016; Ramsey and Pence 2016; Sober 1993; Walsh 2007; Pence and Ramsey 2015; Huneman 2015).

Chapter 3

3.1 The quasispecies concept at the origin of life is a good example of the disconnect between theoretical and empirical studies. On the one hand, the small replicators used in theoretical studies do not have laboratory exemplars. On the other hand, the molecules that are available experimentally are not suitable for testing the hypotheses that were developed theoretically.

3.2 I think it is important from both a historical and ethical perspective to highlight an oft-forgotten feature when reporting on others' works. Any historian of science will remind one that accounts of who discovered what are fraught with problems. Sometimes the credit should not be attributed to one individual because in reality most advances result from a collection of works or exchanges that shaped an idea. Stigler's law of eponymy goes further and states that no scientific discovery is named after the original discoverer (Stigler 1980). That is, perhaps, too strongly worded, but the point is noted. Naming individuals in this book is done for historical reasons or convenience or because their work or writings captured the advance with clarity. It also assists readers in locating some of the most helpful references. Of course, the individual's contribution must not be minimized, but for accuracy and ethics, it should be remembered that several people whose names are lost in history were likely instrumental. I have also taken the liberty of referring to my own works where I think it is appropriate.

This additional note 3.2 is not specific to any particular researcher and should be received as a general point.

Chapter 4

4.1 There have been an enormous number of publications related to the early chemistry of sugars, nucleobases, nucleosides, and nucleotides and the subsequent experimental optimization to increase yields and molecular variants. The references here provide only a glimpse of a vast body of work.

Chapter 5

5.1 The term *complexity* (Adami 2002) can be used in many ways (see also chapter 2). In ribozymes, it refers to the nature of the sequence (length and nucleotide content), structure (folded nature of the molecule), or function (biochemical nature of the catalytic reactions). In this chapter, the term is used loosely, but the general idea is that the increase in size, number of ribonucleotides, and diversity of structural folds correlates positively with our intuitive understanding of complexity in the early ribozymes.

5.2 As discussed in chapters 3 and 6, biochemists and evolutionists use terms differently. For biochemists, cooperation may simply mean that two molecules interact with each other so that their functions are enhanced, or a new function is realized. Hemoglobin monomers form tetramers to bind oxygen with greater affinity. In evolutionary biology the term *cooperation* has special significance with regard to the sociobiology of individuals.

5.3 The functional constraint described in this section is due to the biochemical nature of ribozymes. Simple ligases are small and the accessibility of substrates to the active site is relatively unrestricted. In contrast, the larger, more complex ribozymes are more tightly folded, the active site is less accessible, and steric hindrance leads to greater substrate specificity.

Chapter 6

6.1 It is worth clarifying that for some of the major ETs like the origin of photosynthesis or the evolution of language, the issues concerning the units or levels of selection are not connected to new "kinds" of individuals. This is one of the distinctions between ETs and ETIs. ETs are, of course, still part of the more general debate about units, levels, and the target of natural selection whatever one argues them to be. But in ETs like photosynthesis or the evolution of language there is no transition to a new kind of individual and subsequent change in the level of selection. A discussion of the units and levels of selection requires a dedicated section and is dealt with further in chapter 13. For the field of ETIs, the references included in this chapter will provide the reader with a basic appreciation of ETIs. For a more thorough treatment, especially with regard to the evolution of multicellularity, which is perhaps the premier example of an ETI, see Buss 1987; Michod 1999, 2007; Michod and Nedelcu 2003; Michod and Roze 1997; West et al. 2015.

6.2 There are different classes of MGEs (mobile genetic elements), depending on their mode of reproduction. For a review of the various types and how they impact evolutionary processes see Frost et al. 2005.

6.3 There are two basic components (reproduction and viability) that govern the fitness of any replicating unit, whether it be an MGE, genome, cell, etc. At its most fundamental level, a replicating unit's life history can be described in these two parameters and the trade-off between them. For further reading on life history evolution see Roff 2002, Stearns 1992. The covariance effect is described in Michod 2006, and for a specific example of the covariance effect applied to an MGE see fig. 2 in Durand and Michod 2010. For a more general discussion of covariance, group selection, and multilevel selection theory, see Okasha 2006.

6.4 I have alluded to the importance of terminology several times and it is again necessary to clarify an issue around the term *division of labor*, especially for students and researchers who have a deeper interest in this. As Lloyd has pointed out, terminology and the questions asked using a specific terminology have a profound effect on the interpretations of causality in evolution (Lloyd 2015). The term *division of labor* is a case in point. Some researchers are critical of the term because of the limitations it places on unraveling the causes of functional divergence. It is argued that the term *task allocation* is preferable. (For a discussion of this see Ketcham 2019.) I prefer task allocation but have persisted with division of labor here because of its wider usage.

Chapter 7

7.1 The LRUs comprise structurally and functionally separate molecules, but each molecule is not an autonomous individual. Calling the molecules individuals simply means that they are structurally distinct. The group of molecules (an LRU) can be called an individual in the evolutionary sense, since it is the group that replicates and produces offspring.

PART 2

Chapter 8

8.1 In the evolutionary interpretation of death (as opposed to biochemical interpretations), both components, viability and reproduction, must be extinguished. This is required to separate death from, for example, spores (which retain a *potential* for viability and reproduction, sometimes for hundreds of millions of years [Vreeland, Rosenzweig, and Powers 2000], but show no signs of life), senescence (where the organism stops reproducing but is still, for a while at least, viable), or suspended animation (the brine shrimp example in chapter 2).

8.2 There are many terms used by researchers to contrast PCD with forms of death that do not exhibit a programmed component. They include (i) *death*, which is sometimes used casually as a way of differentiating death from PCD when the biochemical pathways for death have not been activated; (ii) *incidental death*, which often refers to cells that die as a result of physical or chemical damage, for example, exposure to a harmful chemical or mechanical damage; (iii) *lytic death*, which usually refers to situations where a cell lyzes due to mechanical rupture or viral lysis; (iv) *necrosis*, which also refers to a type of lysis, usually not in the context of viral lysis, but rather physical or chemical lysis; and (iv) *non-PCD*, which is sometimes used generically to describe any circumstance when the PCD genetic mechanism plays no part.

Chapter 9

No additional notes.

Chapter 10

10.1 The evolutionary relationships between members of the C14 peptidase family have largely been resolved (see the references to Aravind, Choi, Koonin, Uren, and colleagues). The homologous enzymes in phytoplankton are metacaspases (as opposed to caspases, orthocaspases, or paracaspases) and there now appears to be consensus concerning the terminology (Minina et al. 2020). However, for biochemists, the issue of functionality is still debated. The titles of two references cited here could not be more oppositional: "Metacaspases are caspases. Doubt no more" by Carmona-Gutierrez et al. and "Metacaspases are not caspases—always doubt" by Enoksson and Salvesen. This is discussed in the section on "Measures of PCD in unicellular eukaryotes."

10.2 The sensitivity, specificity, and positive or negative predictive values of any assay can be quantified, but there are insufficient data to do this for the PCD assays and there are no studies specifically examining these parameters. I have relied on literature reviews, experimental data, and qualitative reports to make these judgments.

Chapter 11

11.1 Considering that the cell dies, it seems impossible that PCD can be an adaptation at the level of the cell. However, some researchers have used the term to describe a developmental step in some unicellular organism's life cycle, for example, the liberation of the spore from the mother cell in *B. subtilis* (Decker and Ramamurthi 2017).

Chapter 12

12.1 Co-option was recognized as a process for evolving new functions since the time of Charles Darwin and others (McLennan 2008). It is a way of explaining the emergence of new adaptations that are otherwise difficult to explain using phyletic gradualism where "transitional" forms appear to be at a fitness disadvantage. It should be remembered that co-option is only one of many ways that explain the rapid emergence of new functions (see for example Lin, Kazlauskas, and Travisano 2017). In addition, and as with so many evolutionary terms, the term *co-option* is not always used consistently (McLennan 2008).

12.2 The non-adaptive versus adaptive debate concerning PCD has, for a long time, been the focus of much of PCD evolution. For this reason, I dedicated whole chapters to the two contrasting views. It really does depend on what is meant and how one interprets the terms, definitions, and ecological contexts. In collaboration with Grant Ramsey (Durand and Ramsey 2019), we developed evolutionary definitions for death (true and ersatz PCD, and incidental death) that will hopefully solve some of these issues discussed in chapters 8, 11, and 12).

Chapter 13

13.1 PCD is often interpreted using the unit of selection and the levels of selection. However, the problem is that authors usually do not state what they mean by "units" and "levels," so following their arguments can sometimes be very difficult. There continue to be publications related to the problem of units and levels, but the more recent works are not covered here. One of the older references that newcomers to the field may find helpful is Mayr's perspective from 1997 (Mayr 1997).

13.2 The difference between cooperation and altruism is not always made explicit by some authors. Strictly speaking, in cooperation there is no fitness cost to the actor, whereas in altruism there is a fitness cost associated with the behavior. In both instances, the behavior (cooperative or altruistic) must be selected for because of the benefit it provides.

13.3 Discussing the various decompositions of the Price equation is not necessary for this discussion. The important issue of transmission bias and the PCD trait are mentioned only to make the reader aware that these are not being ignored. Interested readers are referred to the works of Okasha (for discussions of the price equation and MLS1) (Okasha 2006), Luque (for the various decompositions of the Price equation) (Luque 2017), and Durand and Ramsey (for a discussion of PCD and the Price equation) (Durand and Ramsey 2019).

Chapter 14

No additional notes.

PART 3

Chapter 15

15.1 It must be remembered that the term *group* is used differently by biochemists and evolutionists. This is discussed in more detail in chapter 6. Briefly, in the first case, where the phrase "groups of molecules" is used in the biochemical sense, no individual molecule is capable of independent replication. The emergent property (in the biochemical sense) is the replication of the collection of molecules. In the second case, the phrase "groups of LRUs" is used in the evolutionary sense. The LRUs are reproducing individuals that cooperate to form a HRU.

Chapter 16

16.1 Similar interactions are described in many examples (see the references to Vardi, Bidle, Falkowski, Berman-Frank, Bar-Zeev, Bowler, and others). The studies are not always framed in terms of coevolution, but the data and interpretations revealed how two taxa coadapted to the nutritional requirements via a PCD mechanism.

16.2 Artificial selection of colonial groups (simple multicellularity) can lead to stable, heritable phenotypes (Boraas, Seale, and Boxhorn 1998; Ratcliff et al. 2912; Ratcliff et al. 2013; Herron et al. 2019). Artificially selecting for a phenotype usually makes use of strong selection pressures and can expose constraints and trade-offs that result in forms that will be compromised in natural settings (Boyd, Rosenzweig, and Herron 2018).

16.3 Students of the subject note that the terminology can sometimes be confusing. Coloniality is not the same as multicellularity and there are multiple stages between the two. Using the terms interchangeably does not do justice to the theoretical work that suggests a much more layered process as life transitioned from unicellularity through coloniality / aggregation to simple multicellularity and eventually complex (true) multicellularity. Even when we do have definitions the issue is not always clear-cut; see for example discussions of plant genets and ramets and the levels of selection (Okasha 2006; Pineda-Krch and Lehtilä 2004; Clarke 2011), the different "kinds" of individuals (Buss 1987; Lidgard and Nyhart 2017), and multicellularity (Niklas and Newman 2013).

Reference list

Abada, A., and E. Segev. 2018. "Multicellular features of phytoplankton." *Frontiers in Marine Science* 5 (144).

Abbot, P., J. Abe, J. Alcock, S. Alizon, J. A. Alpedrinha, M. Andersson, J. B. Andre, M. van Baalen, F. Balloux, S. Balshine, N. Barton, L. W. Beukeboom, J. M. Biernaskie, T. Bilde, G. Borgia, M. Breed, S. Brown, R. Bshary, A. Buckling, N. T. Burley, M. N. Burton-Chellew, M. A. Cant, M. Chapuisat, E. L. Charnov, T. Clutton-Brock, A. Cockburn, B. J. Cole, N. Colegrave, L. Cosmides, I. D. Couzin, J. A. Coyne, S. Creel, B. Crespi, R. L. Curry, S. R. Dall, T. Day, J. L. Dickinson, L. A. Dugatkin, C. El Mouden, S. T. Emlen, J. Evans, R. Ferriere, J. Field, S. Foitzik, K. Foster, W. A. Foster, C. W. Fox, J. Gadau, S. Gandon, A. Gardner, M. G. Gardner, T. Getty, M. A. Goodisman, A. Grafen, R. Grosberg, C. M. Grozinger, P. H. Gouyon, D. Gwynne, P. H. Harvey, B. J. Hatchwell, J. Heinze, H. Helantera, K. R. Helms, K. Hill, N. Jiricny, R. A. Johnstone, A. Kacelnik, E. T. Kiers, H. Kokko, J. Komdeur, J. Korb, D. Kronauer, R. Kümmerli, L. Lehmann, T. A. Linksvayer, S. Lion, B. Lyon, J. A. Marshall, R. McElreath, Y. Michalakis, R. E. Michod, D. Mock, T. Monnin, R. Montgomerie, A. J. Moore, U. G. Mueller, R. Noë, S. Okasha, P. Pamilo, G. A. Parker, J. S. Pedersen, I. Pen, D. Pfennig, D. C. Queller, D. J. Rankin, S. E. Reece, H. K. Reeve, M. Reuter, G. Roberts, S. K. Robson, D. Roze, F. Rousset, O. Rueppell, J. L. Sachs, L. Santorelli, P. Schmid-Hempel, M. P. Schwarz, T. Scott-Phillips, J. Shellmann-Sherman, P. W. Sherman, D. M. Shuker, J. Smith, J. C. Spagna, B. Strassmann, A. V. Suarez, L. Sundström, M. Taborsky, P. Taylor, G. Thompson, J. Tooby, N. D. Tsutsui, K. Tsuji, S. Turillazzi, F. Ubeda, E. L. Vargo, B. Voelkl, T. Wenseleers, S. A. West, M. J. West-Eberhard, D. F. Westneat, D. C. Wiernasz, G. Wild, R. Wrangham, A. J. Young, D. W. Zeh, J. A. Zeh, and A. Zink. 2011. Inclusive fitness theory and eusociality. *Nature* 471 (7339):E1–E4.

Adami, C. 2002. "What is complexity?" *Bioessays* 24 (12):1085–1094. doi: 10.1002/bies.10192.

Affenzeller, M. J., A. Darehshouri, A. Andosch, C. Lutz, and U. Lutz-Meindl. 2009. "PCD and autophagy in the unicellular green alga *Micrasterias denticulata*." *Autophagy* 5 (6):854–855.

Agren, J. A. 2014. "Evolutionary transitions in individuality: insights from transposable elements." *Trends in Ecology and Evolution* 29 (2):90–96. doi: 10.1016/j.tree.2013.10.007.

Airapetian, V. S., A. Glocer, G. Gronoff, E. Hébrard, and W. Danchi. 2016. "Prebiotic chemistry and atmospheric warming of early Earth by an active young Sun." *Nature Geoscience* 9:452–455.

Aizenman, E., H. Engelberg-Kulka, and G. Glaser. 1996. "An *Escherichia coli* chromosomal 'addiction module' regulated by guanosine [corrected] 3′,5′-bispyrophosphate: a model for programmed bacterial cell death." *Proceedings of the National Academy of Science USA* 93:6059–6063.

Aktipis, C. A., and R. M. Nesse. 2013. "Evolutionary foundations for cancer biology." *Evolutionary Applications* 6:144–159. doi: 10.1111/eva.12034.

Al-Olayan, E. M., G. T. Williams, and H. Hurd. 2002. "Apoptosis in the malaria protozoan, *Plasmodium berghei*: a possible mechanism for limiting intensity of infection in the mosquito." *International Journal of Parasitology* 32 (9):1133–1143.

Ameisen, J. C. 2002. "On the origin, evolution, and nature of programmed cell death: a timeline of four billion years." *Cell Death and Differentiation* 9 (4):367–393. doi: 10.1038/sj/cdd/4400950.

Andrade, M. 1996. "Sexual selection for male sacrifice in the Australian redback spider." *Science* 271:70–72.

Aravind, L., V. M. Dixit, and E. V. Koonin. 1999. "The domains of death: evolution of the apoptosis machinery." *Trends in Biochemical Science* 24 (2):47–53.

Aravind, L., and E. V. Koonin. 2002. "Classification of the caspase-hemoglobinase fold: detection of new families and implications for the origin of the eukaryotic separins." *Proteins* 46 (4):355–367.

Ariew, A. 2002. "Platonic and Aristotelian roots in teleological arguments." In *Functions: new essays in the philosophy of psychology and biology*, edited by A. Ariew, R. Cummins, and M. Perlman. Oxford, UK: Oxford University Press.

Armbrust, E. V., J. A. Berges, C. Bowler, B. R. Green, D. Martinez, N. H. Putnam, S. Zhou, A. E. Allen, K. E. Apt, M. Bechner, M. A. Brzezinski, B. K. Chaal, A. Chiovitti, A. K. Davis, M. S. Demarest, J. C. Detter, T. Glavina, D. Goodstein, M. Z. Hadi, U. Hellsten, M. Hildebrand, B. D. Jenkins, J. Jurka, V. V. Kapitonov, N. Kroger, W. W. Lau, T. W. Lane, F. W. Larimer, J. C. Lippmeier, S. Lucas, M. Medina, A. Montsant, M. Obornik, M. S. Parker, B. Palenik, G. J. Pazour, P. M. Richardson, T. A. Rynearson, M. A. Saito, D. C. Schwartz, K. Thamatrakoln, K. Valentin, A. Vardi, F. P. Wilkerson, and D. S. Rokhsar. 2004. "The genome of the diatom *Thalassiosira pseudonana*: ecology, evolution, and metabolism." *Science* 306 (5693):79–86. doi: 10.1126/science.1101156.

Arnoult, D., I. Tatischeff, J. Estaquier, M. Girard, F. Sureau, J. P. Tissier, A. Grodet, M. Dellinger, F. Traincard, A. Kahn, J. C. Ameisen, and P. X. Petit. 2001. "On the evolutionary conservation of the cell death pathway: mitochondrial release of an apoptosis-inducing factor during *Dictyostelium discoideum* cell death." *Molecular Biology of the Cell* 12 (10):3016–3030.

Bartel, D. P., and J. W. Szostak. 1993. "Isolation of new ribozymes from a large pool of random sequences." *Science* 261 (5127):1411–1418.

Bar-Zeev, E., I. Avishay, K. D. Bidle, and I. Berman-Frank. 2013. "Programmed cell death in the marine cyanobacterium *Trichodesmium* mediates carbon and nitrogen export." *ISME Journal* 7 (12):2340–2348. doi: 10.1038/ismej.2013.121.

Bayles, K. W. 2007. "The biological role of death and lysis in biofilm development." *Nature Reviews Microbiology* 5 (9):721–726. doi: 10.1038/nrmicro1743.

Bayles, K. W. 2014. "Bacterial programmed cell death: making sense of a paradox." *Nature Reviews Microbiology* 12 (1):63–69. doi: 10.1038/nrmicro3136.

Becks, L., S. P. Ellner, L. E. Jones, and N. G. Hairston. 2010. "Reduction of adaptive genetic diversity radically alters eco-evolutionary community dynamics." *Ecology Letters* 13:989–997.

Bell, R. A. V., and L. A. Megeney. 2017. "Evolution of caspase-mediated cell death and differentiation: twins separated at birth." *Cell Death and Differentiation* 24 (8):1359–1368. doi: 10.1038/cdd.2017.37.

Berges, J. A., and C. J. Choi. 2014. "Cell death in algae: physiological processes and relationships with stress." *Perspectives in Phycology* 1:103–112.

Berges, J. A., and P. G. Falkowski. 1998. "Physiological stress and cell death in marine phytoplankton: induction of proteases in response to nitrogen or light limitation." *Limnology and Oceanography* 43:129–135.

Berman-Frank, I., K. D. Bidle, L. Haramaty, and P. G. Falkowski. 2004. "The demise of the marine cyanobacterium, *Trichodesmium* spp., via an autocatalyzed cell death pathway." *Limnology and Oceanography* 49:997–1005.

Bernhardt, H. S. 2012. "The RNA world hypothesis: the worst theory of the early evolution of life (except for all the others)(a)." *Biology Direct* 7:23. doi: 10.1186/1745-6150-7-23.

Bich, L., and S. Green. 2018. "Is defining life pointless? Operational definitions at the frontiers of biology." *Synthese* 195:3919–3946. doi: 10.1007/s11229-017-1397-9.

Bich, L., M. Mossio, K. Ruiz-Mirazo, and A. Moreno. 2015. "Biological regulation: controlling the system from within." *Biology and Philosophy* 31 (2):237–265. doi: 10.1007/s10539-015-9497-8.

Bidle, K. D. 2015. "The molecular ecophysiology of programmed cell death in marine phytoplankton." *Annual Review of Marine Science* 7:341–375.

Bidle, K. D. 2016. "Programmed cell death in unicellular phytoplankton." *Current Biology* 26 (13):R594–R607. doi: 10.1016/j.cub.2016.05.056.

Bidle, K. D., and S. J. Bender. 2008. "Iron starvation and culture age activate metacaspases and programmed cell death in the marine diatom *Thalassiosira pseudonana*." *Eukaryote Cell* 7 (2):223–236. doi: 10.1128/EC.00296-07.

Bidle, K. D., and P. G. Falkowski. 2004. "Cell death in planktonic, photosynthetic microorganisms." *Nature Reviews Microbiology* 2 (8):643–655. doi: 10.1038/nrmicro956.

Bidle, K. A., L. Haramaty, N. Baggett, J. Nannen, and K. D. Bidle. 2010. "Tantalizing evidence for caspase-like protein expression and activity in the cellular stress response of Archaea." *Environmental Microbiology* 12 (5):1161–1172. doi: 10.1111/j.1462-2920.2010.02157.x.

Bidle, K. D., L. Haramaty, J. Barcelos e Ramos, and P. Falkowski. 2007. "Viral activation and recruitment of metacaspases in the unicellular coccolithophore, *Emiliania huxleyi*." *Proceedings of the National Academy of Science USA* 104 (14):6049–6054. doi: 10.1073/pnas.0701240104.

Birch, J. 2014. "Hamilton's rule and its discontents." *British Journal for the Philosophy of Science* 65:381–411.

Birch, J. 2017. *The philosophy of social evolution*. Oxford, UK: Oxford University Press.

Birch, J., and S. Okasha. 2015. "Kin selection and its critics." *BioScience* 65:22–32.

Blackstone, N. W. 2013. "Why did eukaryotes evolve only once? Genetic and energetic aspects of conflict and conflict mediation." *Philosophical Transactions of the Royal Society London B* 368:20120266. doi: 10.1098/rstb.2012.0266.

Blackstone, N. W. 2016. "An evolutionary framework for understanding the origin of eukaryotes." *Biology (Basel)* 5 (2):E18. doi: 10.3390/biology5020018.

Blackstone, N. W., and D. R. Green. 1999. "The evolution of a mechanism of cell suicide." *Bioessays* 21:84–88. doi: 10.1002/(SICI)1521-1878(199901)21:1<84::AID-BIES11>3.0.CO;2-0.

Blouin, M. S. 2003. "DNA-based methods for pedigree reconstruction and kinship analysis in natural populations." *Trends in Ecology and Evolution* 18:503–511.

Bludau, I., and R. Aebersold. 2020. "Proteomic and interactomic insights into the molecular basis of cell functional diversity." *Nature Reviews Molecular Cell Biology*. In press, doi.org/10.1038/s41580-020-0231-2.

Bohler, C., P. E. Nielsen, and L. E. Orgel. 1995. "Template switching between PNA and RNA oligonucleotides." *Nature* 376:578–581. doi: 10.1038/376578a0.

Bonduriansky, R., and T. Day. 2018. *Extended heredity: a new understanding of inheritance and evolution*. Princeton, USA: Princeton University Press.

Boraas, M. E., D. B. Seale, and J. E. Boxhorn. 1998. "Phagotrophy by a flagellate selects for colonial prey: a possible origin of multicellularity." *Evolutionary Ecology* 12:153–164.

Borrello, M. E. 2005. "The rise, fall and resurrection of group selection." *Endeavour* 29 (1):43–47. doi: 10.1016/j.endeavour.2004.11.003.

Bouderbala, I., N. El Saadi, A. Bah, and P. Auger. 2018. "A 3D individual-based model to study effects of chemotaxis, competition and diffusion on the motile-phytoplankton aggregation." *Acta Biotheoretica* 66:257–278. doi: 10.1007/s10441-018-9318-y.

Bouderbala, I., N. El Saadi, A. Bah, and P. Auger. 2019. "A simulation study on how the resource competition and anti-predator cooperation impact the motile-phytoplankton groups' formation under predation stress." *Ecological Modelling* 391:16–28.

Bourrat, P. 2015. "Levels, time and fitness in evolutionary transitions in individuality." *Philosophical and Theoretical Biology* 7 (e601).

Boyd, M., F. Rosenzweig, and M. D. Herron. 2018. "Analysis of motility in multicellular *Chlamydomonas reinhardtii* evolved under predation." *PLOS ONE* 13 (1):e0192184. doi: 10.1371/journal.pone.0192184.

Boyle, E. A., Y. I. Li, and J. K. Pritchard. 2017. "An expanded view of complex traits: from polygenic to omnigenic." *Cell* 169 (7):1177–1186. doi: 10.1016/j.cell.2017.05.038.

Brandon, R. 1990. *Adaptation and environment*. Princeton, USA: Princeton University Press.

Brandon, R., and S. Carson. 1996. "The indeterministic character of evolutionary change." *Philosophy of Science* 63:315–337.

Brosius, J. 1999. "Genomes were forged by massive bombardments with retroelements and retrosequences." *Genetica* 107 (1–3):209–238.

Burt, A., and R. Trivers. 2006. *Genes in conflict: the biology of selfish genetic elements*. Cambridge, USA: Belknap Press of Harvard University Press.

Buss, L. W. 1987. *The evolution of individuality*. Princeton, USA: Princeton University Press.

Canil, D. 2002. "Vanadium in peridotites, mantle redox and tectonic environments: Archean to present." *Earth and Planetary Science Letters* 195:75–90.

Carmona, D., C. R. Fitzpatrick, and M. T. Johnson. 2015. "Fifty years of co-evolution and beyond: integrating co-evolution from molecules to species." *Molecular Ecology* 24 (21):5315–5329. doi: 10.1111/mec.13389.

Carmona-Gutierrez, D., M. A. Bauer, A. Zimmermann, A. Aguilera, N. Austriaco, K. Ayscough, R. Balzan, S. Bar-Nun, A. Barrientos, P. Belenky, M. Blondel, R. J. Braun, M. Breitenbach, W. C. Burhans, S. Buttner, D. Cavalieri, M. Chang, K. F. Cooper, M. Corte-Real, V. Costa, C. Cullin, I. Dawes, J. Dengjel, M. B. Dickman, T. Eisenberg, B. Fahrenkrog, N. Fasel, K. U. Frohlich, A. Gargouri, S. Giannattasio, P. Goffrini, C. W. Gourlay, C. M. Grant, M. T. Greenwood, N. Guaragnella, T. Heger, J. Heinisch, E. Herker, J. M. Herrmann, S. Hofer, A. Jimenez-Ruiz, H. Jungwirth, K. Kainz, D. P. Kontoyiannis, P. Ludovico, S. Manon, E. Martegani, C. Mazzoni, L. A. Megeney, C. Meisinger, J. Nielsen, T. Nystrom, H. D. Osiewacz, T. F. Outeiro, H. O. Park, T. Pendl, D. Petranovic, S. Picot, P. Polcic, T. Powers, M. Ramsdale, M. Rinnerthaler, P. Rockenfeller, C. Ruckenstuhl, R. Schaffrath, M. Segovia, F. F. Severin, A. Sharon, S. J. Sigrist, C. Sommer-Ruck, M. J. Sousa, J. M. Thevelein, K. Thevissen, V. Titorenko, M. B. Toledano, M. Tuite, F. N. Vogtle, B. Westermann, J. Winderickx, S. Wissing, S. Wolfl, Z. J. Zhang, R. Y. Zhao, B. Zhou, L. Galluzzi, G. Kroemer, and F. Madeo. 2018. "Guidelines and recommendations on yeast cell death nomenclature." *Microbial Cell* 5 (1):4–31. doi: 10.15698/mic2018.01.607.

Carmona-Gutierrez, D., T. Eisenberg, S. Buttner, C. Meisinger, G. Kroemer, and F. Madeo. 2010. "Apoptosis in yeast: triggers, pathways, subroutines." *Cell Death and Differentiation* 17 (5):763–773. doi: 10.1038/cdd.2009.219.

Carmona-Gutierrez, D., K. U. Frohlich, G. Kroemer, and F. Madeo. 2010. "Metacaspases are caspases. Doubt no more." *Cell Death and Differentiation* 17 (3):377–378. doi: 10.1038/cdd.2009.198.

Carter, C. W., Jr., and R. Wolfenden. 2015. "tRNA acceptor stem and anticodon bases form independent codes related to protein folding." *Proceedings of the National Academy of Science USA* 112 (24):7489–7494. doi: 10.1073/pnas.1507569112.

Casiraghi, M., A. Galimberti, A. Sandionigi, A. Bruno, and M. Labra. 2016. "Life with or without names." *Evolutionary Biology* 43:582–595.

Cebollero, E., and F. Reggiori. 2009. "Regulation of autophagy in yeast *Saccharomyces cerevisiae*." *Biochimica et Biophysica Acta* 1793 (9):1413–1421. doi: 10.1016/j.bbamcr.2009.01.008.

Chandramohan, L., J. S. Ahn, K. E. Weaver, and K. W. Bayles. 2009. "An overlap between the control of programmed cell death in *Bacillus anthracis* and sporulation." *Journal of Bacteriology* 191 (13):4103–4310. doi: 10.1128/JB.00314-09.

Chen, W. X., X. M. Liu, M. M. Lv, L. Chen, J. H. Zhao, S. L. Zhong, M. H. Ji, Q. Hu, Z. Luo, J. Z. Wu, and J. H. Tang. 2014. "Exosomes from drug-resistant breast cancer cells transmit chemoresistance by a horizontal transfer of microRNAs." *PLOS ONE* 9 (4):e95240. doi: 10.1371/journal.pone.0095240.

Choi, C. J., and J. A. Berges. 2013. "New types of metacaspases in phytoplankton reveal diverse origins of cell death proteases." *Cell Death and Disease* 4:e490. doi: 10.1038/cddis.2013.21.

Cilliers, P. 1998. *Complexity and postmodernism: understanding complex systems*. London, UK: Routledge.

Clarke, E. 2011. "Plant individuality and multilevel selection theory." In: *The major transitions in evolution revisited*, edited by B. Calcott and K. Sterelny, 227–250. Boston, USA: MIT Press.

Collin, R. 1906. "Recherches cytologiques sur le développment de la cellule nerveuse." *Névrae* 8:181–308.

Cornillon, S., C. Foa, J. Davoust, N. Buonavista, J. D. Gross, and P. Golstein. 1994. "Programmed cell death in *Dictyostelium*." *Journal of Cell Science* 107:2691–2704.

Daly, T., X. Chen, and D. Penny. 2011. "How old are RNA networks?" In *RNA infrastructure and networks*, edited by L. J. Collins, 255–273. Austin, USA: Landes Bioscience.

Damuth, J., and I. L. Heisler. 1988. "Alternate formulations of multilevel selection." *Biology and Philosophy* 3:407–430.

Darwin, C. 1859. *On the origin of species by means of natural selection, or the preservation of favoured races in the struggle for life*. London, UK: John Murray.

Darwin, C. 1862. *On the various contrivances by which British and foreign orchids are fertilized by insects and on the good effects of intercrossing*. London, UK: John Murray.

Darwin, C., and A. Wallace. 1858. "On the tendency of species to form varieties; and on the perpetuation of varieties and species by natural means of selection." *Journal of the Proceedings of the Linnean Society: Zoology* 3:45–62.

Daugherty, M. D., and H. S. Malik. 2012. "Rules of engagement: molecular insights from host-virus arms races." *Annual Review of Genetics* 46:677–700. doi: 10.1146/annurev-genet-110711-155522.

Dawkins, R. 1982. *The extended phenotype: the gene as the unit of selection*. San Francisco, USA: Freeman.

Dawkins, R. 2006. *The selfish gene*. 30th anniversary ed. Oxford, UK: Oxford University Press.

Debrabant, A., and H. Nakhasi. 2003. "Programmed cell death in Trypanosomatids: is it an altruistic mechanism for survival of the fittest?" *Kinetoplastid Biolology and Disease* 2:7. doi: 10.1186/1475-9292-2-7.

Deck, C., M. Jauker, and C. Richert. 2011. "Efficient enzyme-free copying of all four nucleobases templated by immobilized RNA." *Nature Chemistry* 3:603–608. doi: 10.1038/nchem.1086.

Decker, A. R., and K. S. Ramamurthi. 2017. "Cell death pathway that monitors spore morphogenesis." *Trends in Microbiology* 25 (8):637–647. doi: 10.1016/j.tim.2017.03.005.

Delehanty, M. 2005. "Emergent properties and the context objection to reduction." *Biology and Philosophy* 20 (4):715–734.

Denny, M., and S. Gaines. 2002. *Chance in biology: using probability to explore nature*. Princeton, USA: Princeton University Press.

Deplazes, A., and M. Huppenbauer. 2009. "Synthetic organisms and living machines: positioning the products of synthetic biology at the borderline between living and non-living matter." *Systems and Synthetic Biology* 3 (1–4):55–63. doi: 10.1007/s11693-009-9029-4.

Deponte, M. 2008. "Programmed cell death in protists." *Biochimica et Biophysica Acta* 1783 (7):1396–1405. doi: 10.1016/j.bbamcr.2008.01.018.

Descartes, R. 1642. *Meditationes de prima philosophia, in quibus Dei existentia & animae humanae à corpore distinctio demonstrantur: his adjunctae sunt variae objectiones doctorum virorum in istas de Deo & anima demonstrationes, cum responsionibus authoris*. Amsterdam, Netherlands: Elzivir.

Dhar, N., M. S. Weinberg, R. E. Michod, and P. M. Durand. 2017. "Molecular trade-offs in RNA ligases affected the modular emergence of complex ribozymes at the origin of life." *Royal Society Open Science* 4 (9):170376. doi: 10.1098/rsos.170376.

Dieckmann, U., and M. Doebeli. 2005. "Pluralism in evolutionary theory." *Journal of Evolutionary Biology* 18:1209–1213.

Ding, Y., N. Gan, J. Li, B. Sedmak, and L. Song. 2012. "Hydrogen peroxide induces apoptotic-like cell death in *Microcystis aeruginosa* (Chroococcales, Cyanobacteria) in a dose-dependent manner." *Phycologia* 51:567–575.

Dingman, J. E., and J. E. Lawrence. 2012. "Heat-stress-induced programmed cell death in *Heterosigma akashiwo* (Raphidophyceae)." *Harmful Algae* 16:108–116.

Dominguez, F., and F. J. Cejudo. 2014. "Programmed cell death (PCD): an essential process of cereal seed development and germination." *Frontiers in Plant Science* 5. doi: ARTN 36610.3389/fpls.2014.00366.

Draganić, I. G. 2005. "Radiolysis of water: a look at its origin and occurrence in the nature." *Radiation Physics and Chemistry* 72:181–186.

Drenkard, S., J. Ferris, and A. Eschenmoser. 1990. "Chemistry of alpha-aminonitriles. Aziridine-2-carbonitrile: photochemical formation from 2-aminopropenenitrile." *Helvetica Chimica Acta* 73:1373–1390. doi: 10.1002/hlca.19900730524.

Dugatkin, L. A. 1997. *Cooperation among animals: an evolutionary perspective.* Oxford Series in Ecology and Evolution. Oxford, UK: Oxford University Press.

Dupré, J. 2010. "It is not possible to reduce biological explanations to explanations in chemistry and/or physics." In *Contemporary Debates in Philosophy of Biology*, edited by F. J. Ayala and R. Arp, 32–49. Chichester, UK: Wiley-Blackwell.

Durand, P. M., M. M. Barreto Filho, and R. E. Michod. 2019. "Cell death in evolutionary transitions in individuality." *Yale Journal of Biology and Medicine* 92:651–662.

Durand, P. M., R. Choudhury, A. Rashidi, and R. E. Michod. 2014. "Programmed death in a unicellular organism has species-specific fitness effects." *Biology Letters* 10 (2):20131088. doi: 10.1098/rsbl.2013.1088.

Durand, P. M., and R. E. Michod. 2010. "Genomics in the light of evolutionary transitions." *Evolution* 64 (6):1533–1540. doi: 10.1111/j.1558-5646.2009.00907.x.

Durand, P. M., and R. E. Michod. 2011. "What is life and why, how and when did it begin?" *Journal of Cosmology* 64:1533–1540.

Durand, P. M., and G. Ramsey. 2019. "The nature of programmed cell death." *Biological Theory* 14:30–41.

Durand, P. M., A. Rashidi, and R. E. Michod. 2011. "How an organism dies affects the fitness of its neighbors." *American Naturalist* 177:224–232. doi: 10.1086/657686.

Durand, P. M., S. Sym, and R. E. Michod. 2016. "Programmed cell death and complexity in microbial systems." *Current Biology* 26:R587–R593. doi: 10.1016/j.cub.2016.05.057.

Dyrka, W., V. Coustou, A. Daskalov, A. Lends, T. Bardin, M. Berbon, B. Kauffmann, C. Blancard, B. Salin, A. Loquet, and S. J. Saupe. 2020. "Identification of NLR-associated amyloid signaling motifs in filamentous bacteria." bioRxiv preprint doi: https://doi.org/10.1101/2020.01.06.895854.

Eckhart, L., F. Ehrlich, and E. Tschachler. 2019. "A stress response program at the origin of evolutionary innovation in the skin." *Evolutionary Bioinformatics Online* 15:1176934319862246. doi: 10.1177/1176934319862246.

Ehrlich, P. R., and P. H. Raven. 1964. "Butterflies and plants: a study in coevolution." *Evolution* 18:586–608.

Eigen, M. 1971. "Selforganization of matter and the evolution of biological macromolecules." *Naturwissenschaften* 58 (10):465–523.

Eigen, M., and P. Schuster. 1977. "The hypercycle. A principle of natural self-organization. Part A: emergence of the hypercycle." *Naturwissenschaften* 64 (11):541–565.

Eisenberg-Lerner, A., S. Bialik, H.-U. Simon, and A. Kimchi. 2009. "Life and death partners: apoptosis, autophagy and the cross-talk between them." *Cell Death and Differentiation* 16:966–975.

Eldakar, O. T., and D. S. Wilson. 2011. "Eight criticisms not to make about group selection." *Evolution* 65 (6):1523–1526. doi: 10.1111/j.1558-5646.2011.01290.x.

Eme, L., A. Spang, J. Lombard, C. W. Stairs, and T. J. G. Ettema. 2018. "Archaea and the origin of eukaryotes." *Nature Reviews Microbiology* 16 (2):120. doi: 10.1038/nrmicro.2017.154.

Engelberg-Kulka, H., S. Amitai, I. Kolodkin-Gal, and R. Hazan. 2006. "Bacterial programmed cell death and multicellular behavior in bacteria." *PLOS Genetics* 2 (10):e135. doi: 10.1371/journal.pgen.0020135.

Engelberg-Kulka, H., R. Hazan, and S. Amitai. 2005. "mazEF: a chromosomal toxin-antitoxin module that triggers programmed cell death in bacteria." *Journal of Cell Science* 118 (Pt 19):4327–4332. doi: 10.1242/jcs.02619.

Engelbrecht, D., and T. L. Coetzer. 2013. "Turning up the heat: heat stress induces markers of programmed cell death in *Plasmodium falciparum* in vitro." *Cell Death and Disease* 4:e971. doi: 10.1038/cddis.2013.505.

Engelbrecht, D., P. M. Durand, and T. L. Coetzer. 2012. "On programmed cell death in *Plasmodium falciparum*: status quo." *Journal of Tropical Medicine* 2012:646534. doi: 10.1155/2012/646534.

Engelhart, A. E., M. W. Powner, and J. W. Szostak. 2013. "Functional RNAs exhibit tolerance for non-heritable 2′-5′ versus 3′-5′ backbone heterogeneity." *Nature Chemistry* 5 (5):390–394. doi: 10.1038/nchem.1623.

Enoksson, M., and G. S. Salvesen. 2010. "Metacaspases are not caspases—always doubt." *Cell Death and Differentiation* 17 (8):1221. doi: 10.1038/cdd.2010.45.

Erkenbrack, E. M., J. D. Maziarz, O. W. Griffith, C. Liang, A. R. Chavan, M. C. Nnamani, and G. P. Wagner. 2018. "The mammalian decidual cell evolved from a cellular stress response." *PLOS Biology* 16 (8):e2005594. doi: 10.1371/journal.pbio.2005594.

Ernst, M. 1926. "Über Untergang von Zellen während der normalen Entwicklung bei Wirbeltieren." *Z Anat Entwicklungsgesch* 79:228–262.

Fabrizio, P., L. Battistella, R. Vardavas, C. Gattazzo, L. L. Liou, A. Diaspro, J. W. Dossen, E. B. Gralla, and V. D. Longo. 2004. "Superoxide is a mediator of an altruistic aging program in *Saccharomyces cerevisiae*." *Journal of Cell Biology* 166 (7):1055–1067. doi: 10.1083/jcb.200404002.

Fang, F. C., and A. Casadevall. 2011. "Reductionistic and holistic science." *Infection and Immunity* 79 (4):1401–1404. doi: 10.1128/IAI.01343-10.

Farley, J. 1986. "Philosophical and historical aspects of the origin of life." *Treb Soc Cat Biol* 39:37–47.

Fenchel, T. 2008. "The microbial loop–25 years later." *Journal of Experimental Marine Biology and Ecology* 366 (1):99–103.

Ferris, J. P. 2002. "Montmorillonite catalysis of 30–50 mer oligonucleotides: laboratory demonstration of potential steps in the origin of the RNA world." *Origin of Life and Evolution of the Biosphere* 32 (4):311–332.

Ferris, J. P., and G. Ertem. 1993. "Montmorillonite catalysis of RNA oligomer formation in aqueous solution: a model for the prebiotic formation of RNA." *Journal of the American Chemical Society* 115 (26):12270–12275.

Ferris, J. P., A. R. Hill, Jr., R. Liu, and L. E. Orgel. 1996. "Synthesis of long prebiotic oligomers on mineral surfaces." *Nature* 381 (6577):59–61. doi: 10.1038/381059a0.

Field, C. B., M. J. Behrenfeld, J. T. Randerson, and P. Falkowski. 1998. "Primary production of the biosphere: integrating terrestrial and oceanic components." *Science* 281 (5374):237–240.

Filippova, S. N., and K. A. Vinogradova. 2017. "Programmed cell death as one of the stages of streptomycete differentiation." *Microbiology* 86 (4):439–454. doi: 10.1134/S0026261717040075.

Finley, M. I. 1963. *The ancient Greeks: an introduction to their life and thought*. Vol. 20, chapter 6. New York, USA: Viking Press.

Fortunato, A., J. E. Strassmann, L. Santorelli, and D. C. Queller. 2003. "Co-occurrence in nature of different clones of the social amoeba, *Dictyostelium discoideum*." *Molecular Ecology* 12:1031–1038.

Fox Keller, E. 2010. "It is possible to reduce biological explanations to explanations in chemistry and/or physics." In *Contemporary Debates in Philosophy of Biology*, edited by F. J. Ayala and R. Arp, 19–32. Chichester, UK: Wiley-Blackwell.

Foyer, C. H. 2018. "Reactive oxygen species, oxidative signaling and the regulation of photosynthesis." *Environmental and Experimental Botany* 154:134–142. doi: 10.1016/j.envexpbot.2018.05.003.

Frade, J. M., and T. M. Michaelidis. 1997. "Origin of eukaryotic programmed cell death: a consequence of aerobic metabolism?" *Bioessays* 19 (9):827–832. doi: 10.1002/bies.950190913.

Frank, S. A. 1998. *Foundations of social evolution*. Princeton, USA: Princeton University Press.

Frank, S. A. 2012. "Natural selection, IV: the Price equation." *Journal of Evolutionary Biology* 25 (6):1002–1019. doi: 10.1111/j.1420-9101.2012.02498.x.

Franklin, D. J., C. P. D. Brussaard, and J. A. Berges. 2006. "What is the nature and role of programmed cell death in phytoplankton ecology?" *European Journal of Phycology* 41:1–14.

Frost, L. S., R. Leplae, A. O. Summers, and A. Toussaint. 2005. "Mobile genetic elements: the agents of open source evolution." *Nature Reviews Microbiology* 3 (9):722–732. doi: 10.1038/nrmicro1235.

Frost, D. J., U. Mann, Y. Asahara, and D. C. Rubie. 2008. "The redox state of the mantle during and just after core formation." *Philosophical Transactions A Math Phys Eng Sci* 366 (1883):4315–4337. doi: 10.1098/rsta.2008.0147.

Fuller, W. D., R. A. Sanchez, and L. E. Orgel. 1972. "Studies in prebiotic synthesis, VI: synthesis of purine nucleosides." *Journal of Molecular Biology* 67 (1):25–33.

Gánti, T. 1975. "Organization of chemical reactions into dividing and metabolizing units: the chemotons." *Biosystems* 7 (1):15–21.

Gánti, T. 2003a. *Chemoton theory. Vol. 1: Theoretical foundations of fluid machneries. Vol. 2: Theory of living systems*. New York, USA: Kluwer Academic.

Gánti, T. 2003b. *The principles of life*. Oxford, UK: Oxford University Press.

Gao, J., S. Chau, F. Chowdhury, T. Zhou, S. Hossain, G. A. McQuibban, and M. D. Meneghini. 2019. "Meiotic viral attenuation through an ancestral apoptotic pathway." *Proceedings of the National Academy of Science USA* 116:16454–16462. doi: 10.1073/pnas.1900751116.

García-Gómez, C., M. Teresa Mata, F. Van Breusegem, and M. Segovia. 2016. Low-steady-state metabolism induced by elevated CO_2 increases resilience to UV radiation in the unicellular green-algae *Dunaliella tertiolecta*. *Environmental and Experimental Botany* 132:163–174.

Gardner, A. 2017. "The purpose of adaptation." *Interface Focus* 7 (5):20170005. doi: 10.1098/rsfs.2017.0005.

Gardner, A., S. A. West, and G. Wild. 2011. "The genetical theory of kin selection." *Journal of Evolutionary Biology* 24 (5):1020–1043. doi: 10.1111/j.1420-9101.2011.02236.x.

Gayon, J., C. Malaterre, M. Morange, F. Raulin-Cerceau, and S. Tirard. 2010. "Defining Life: conference proceedings." *Origin of Life and Evolution of the Biosphere* 40 (2):119–120. doi: 10.1007/s11084-010-9189-y.

Gilbert, O. M., K. R. Foster, N. J. Mehdiabadi, J. E. Strassmann, and D. C. Queller. 2007. "High relatedness maintains multicellular cooperation in a social amoeba by controlling cheater mutants." *Proceedings of the National Academy of Science USA* 104:8913–8917.

Giroud, C., and W. Eichenberger. 1988. "Fatty-acids of *Chlamydomonas reinhardtii*—structure, positional distribution and biosynthesis." *Biological Chemistry* 369:18–19.

Glücksmann, A. 1951. "Cell deaths in normal vertebrate ontogeny." *Biology Reviews of the Cambridge Philosophical Society* 26 (1):59–86.

Gontier, N. 2015. "Uniting micro- with macroevolution into an extended synthesis: reintegrating life's natural history into evolution studies." In *Macroevolution*, edited by E. Serrelli and N. Gontier. Interdisciplinary Evolution Research. Vol. 2. Switzerland: Springer.

Goodnight, C. J. 1990a. "Experimental studies of community evolution, II: the ecological basis of the response to community selection." *Evolution* 44 (6):1625–1636. doi: 10.1111/j.1558-5646.1990.tb03851.x.

Goodnight, C. J. 1990b. "Experimental studies of community evolution, I: the response to selection at the community level." *Evolution* 44 (6):1614–1624. doi: 10.1111/j.1558-5646.1990.tb03850.x.

Goodnight, C. 2013a. "On multilevel selection and kin selection: contextual analysis meets direct fitness." *Evolution* 67 (6):1539–1548. doi: 10.1111/j.1558-5646.2012.01821.x.

Goodnight, C. J. 2013b. "Defining the individual." In *From groups to individuals*, edited by F. Bouchard and P. Huneman, 37–54. Cambridge, USA: MIT Press.

Gould, S. J. 2002. *The structure of evolutionary theory*. Cambridge, USA: Belknap Press of Harvard University Press.

Gould, S. J., and N. Eldredge. 1977. "Punctuated equilibria: the tempo and mode of evolution reconsidered." *Paleobiology* 3:115–151.

Gould, S. J., and R. C. Lewontin. 1979. "The spandrels of San Marco and the Panglossian paradigm: a critique of the adaptationist programme." *Proceedings of the Royal Society London B* 205 (1161):581–598.

Gould, S. J., and E. A. Lloyd. 1999. "Individuality and adaptation across levels of selection: how shall we name and generalize the unit of Darwinism?" *Proceedings of the National Academy of Science USA* 96 (21):11904–11909.

Gould, S. J., and E. S. Vrba. 1982. "Exaptation—a missing term in the science of form." *Paleobiology* 8:4–15.

Griesemer, J. 2015. "The enduring value of Gánti's chemoton model and life criteria: heuristic pursuit of exact theoretical biology." *Journal of Theoretical Biology* 381:23–28. doi: 10.1016/j.jtbi.2015.05.016.

Griffiths, P. E., and E. M. Neumann-Held. 1999. "The many faces of the gene." *BioScience* 49:656–662.

Gruenheit, N., K. Parkinson, B. Stewart, J. A. Howie, J. B. Wolf, and C. R. L. Thompson. 2017. "A polychromatic 'greenbeard' locus determines patterns of cooperation in a social amoeba." *Nature Communications* 8.

Guseva, E., R. N. Zuckermann, and K. Dill. 2017. "Foldamer hypothesis for the growth and sequence differentiation of prebiotic polymers." *Proceedings of the National Academy of Science USA* 114 (36):E7460–E7468. doi: 10.1073/pnas.1620179114.

Hamburger, V., and R. Levi-Montalcini. 1949. "Proliferation, differentiation and degeneration in the spinal ganglia of the chick embryo under normal and experimental conditions." *Journal of Experimental Zoology* 111:457–501.

Hamilton, W. D. 1964. "The genetical evolution of social behaviour, I and II." *Journal of Theoretical Biology* 7 (1):1–52.

Hanschen, E., D. R. Davison, Z. Grochau-Wright, and R. E. Michod. 2018. "Individuality and the major evolutionary transitions." In *Landscapes of collectivity in the life sciences*, edited by S. Gissis, E. Lamm, and A. Shavit, 255–267. Cambridge, USA: MIT Press.

Hanschen, E. R., D. E. Shelton, and R. E. Michod. 2015. "Evolutionary transitions in individuality and recent models of multicellularity." In *Evolutionary transitions to multicellular life: Principles and mechanisms*, edited by I. Ruiz-Trillo and A. M. Nedelcu, 165–188. Dordrecht: Springer.

Hayden, E. J., and N. Lehman. 2006. "Self-assembly of a group I intron from inactive oligonucleotide fragments." *Chemical Biology* 13 (8):909–918. doi: 10.1016/j.chembiol.2006.06.014.

Hazan, R., and H. Engelberg-Kulka. 2004. "*Escherichia coli* mazEF-mediated cell death as a defense mechanism that inhibits the spread of phage P1." *Molecular Genetics and Genomics* 272 (2):227–234. doi: 10.1007/s00438-004-1048-y.

Hazan, R., B. Sat, and H. Engelberg-Kulka. 2004. "*Escherichia coli* mazEF-mediated cell death is triggered by various stressful conditions." *Journal of Bacteriology* 186 (11):3663–3669. doi: 10.1128/JB.186.11.3663-3669.2004.

Heffernan, G. 1990. *Meditations on First Philosophy/ Meditationes de prima philosophia: A Bilingual Edition (English and Latin Edition)*. Notre Dame, USA: University of Notre Dame Press.

Heisler, I. L., and J. Damuth. 1987. "A method for analyzing selection in hierarchically structured populations." *American Naturalist* 130:582–602.

Helms, M. J., A. Ambit, P. Appleton, L. Tetley, G. H. Coombs, and J. C. Mottram. 2006. "Bloodstream form *Trypanosoma brucei* depend upon multiple metacaspases associated with RAB11-positive endosomes." *Journal of Cell Science* 119 (Pt 6):1105–1117. doi: 10.1242/jcs.02809.

Hendry, A. P., and A. Gonzalez. 2008. "Whither adaptation?" *Biology and Philosophy* 23:673–699.

Heng, H. H. 2009. "The genome-centric concept: resynthesis of evolutionary theory." *Bioessays* 31 (5):512–525. doi: 10.1002/bies.200800182.

Heng, H. H., J. B. Stevens, S. W. Bremer, K. J. Ye, G. Liu, and C. J. Ye. 2010. "The evolutionary mechanism of cancer." *Journal of Cell Biochemistry* 109 (6):1072–1084. doi: 10.1002/jcb.22497.

Herker, E., H. Jungwirth, K. A. Lehmann, C. Maldener, K. U. Frohlich, S. Wissing, S. Buttner, M. Fehr, S. Sigrist, and F. Madeo. 2004. "Chronological aging leads to apoptosis in yeast." *Journal of Cell Biology* 164 (4):501–507. doi: 10.1083/jcb.200310014.

Herre, E. A., and W. T. Wcislo. 2011. "In defence of inclusive fitness theory." *Nature* 471 (7339):E8–9; author reply E9–10. doi: 10.1038/nature09835.

Herron, M. D., J. M. Borin, J. C. Boswell, J. Walker, I.-C. K. Chen, C. A. Knox, M. Boyd, F. Rosenzweig, and W. C. Ratcliff. 2019. "De novo origins of multicellularity in response to predation". *Scientific Reports* 9:2328.

Hey, J. 2006. "On the failure of modern species concepts." *Trends in Ecology Evolution* 21 (8):447–450. doi: 10.1016/j.tree.2006.05.011.

Higgs, P. G., and N. Lehman. 2015. "The RNA world: molecular cooperation at the origins of life." *Nature Reviews Genetics* 16 (1):7–17. doi: 10.1038/nrg3841.

Hochberg, M. E., D. J. Rankin, and M. Taborsky. 2008. "The coevolution of cooperation and dispersal in social groups and its implications for the emergence of multicellularity." *BMC Evolutionary Biology* 8:238. doi:10.1186/1471-2148-8-238.

Hogeweg, P., and N. Takeuchi. 2003. "Multilevel selection in models of prebiotic evolution: compartments and spatial self-organization." *Origin of Life and Evolution of the Biosphere* 33 (4–5):375–403.

Huang, X., B. Huang, J. Chen, and X. Liu. 2016. "Cellular responses of the dinoflagellate *Prorocentrum donghaiense* Lu to phosphate limitation and chronological ageing." *Journal of Plankton Research* 38:83–93.

Huettenbrenner, S., S. Maier, C. Leisser, D. Polgar, S. Strasser, M. Grusch, and G. Krupitza. 2003. "The evolution of cell death programs as prerequisites of multicellularity." *Mutation Research* 543 (3):235–249.

Hull, D. L. 1974. *Philosophy of biological science*. Englewood Cliffs, USA: Prentice-Hall.

Hull, D. L. 1980. "Individuality and selection." *Annual Review of Ecology and Systematics* 11:311–332.

Humphreys, P. W. 1994. "How properties emerge." *Philosophy of Science* 64 (1):1–17.

Humphreys, R., and G. D. Ruxton. 2019. "Adaptive suicide: is a kin-selected driver of fatal behaviours likely?" *Biology Letters* 15:20180823.

Huneman, P., ed. 2013. *Functions: selection and mechanisms*. Dordrecht, Netherlands: Springer.

Huneman, P. 2015. "Evolutionary theory in philosophical focus." In *Handbook of Paleoanthropology*, edited by W. Henke and I. Tattersall, 127–175. Berlin: Springer.

Inoue, T., and L. E. Orgel. 1983. "A nonenzymatic RNA polymerase model." *Science* 219 (4586):859–862.

Iranzo, J., A. E. Lobkovsky, Y. I. Wolf, and E. V. Koonin. 2014. "Virus-host arms race at the joint origin of multicellularity and programmed cell death." *Cell Cycle* 13 (19):3083–3088. doi: 10.4161/15384101.2014.949496.

Jablonski, D. 2008. "Species selection: theory and data." *Annual Review of Ecology, Evolution and Systematics* 39:501–524.

Jacobeen, S., E. C. Graba, C. G. Brandys, T. C. Day, W. C. Ratcliff, and P. J. Yunker. 2018. "Geometry, packing, and evolutionary paths to increased multicellular size." *Physical Review E* 97 (5–1):050401. doi: 10.1103/PhysRevE.97.050401.

James, K. D., and A. D. Ellington. 1999. "The fidelity of template-directed oligonucleotide ligation and the inevitability of polymerase function." *Origins of Life and Evolution of Biospheres* 29 (4):375–390.

Jauzein, C., and D. L. Erdner. 2013. "Stress-related responses in *Alexandrium tama-*

rense cells exposed to environmental changes." *Journal of Eukaryote Microbiology* 60 (5):526–538. doi: 10.1111/jeu.12065.

Jensen, R. B., and K. Gerdes. 1995. "Programmed cell death in bacteria: proteic plasmid stabilization systems." *Molecular Microbiology* 17 (2):205–210.

Jeuken, M. 1975. "The biological and philosophical definitions of life." *Acta Biotheoretica* 24 (1–2):14–21.

Jiang, Q., S. Qin, and Q. Y. Wu. 2010. "Genome-wide comparative analysis of metacaspases in unicellular and filamentous cyanobacteria." *BMC Genomics* 11:198. doi: 10.1186/1471-2164-11-198.

Jimenez, C., J. M. Capasso, C. L. Edelstein, C. J. Rivard, S. Lucia, S. Breusegem, T. Berl, and M. Segovia. 2009. "Different ways to die: cell death modes of the unicellular chlorophyte *Dunaliella viridis* exposed to various environmental stresses are mediated by the caspase-like activity DEVDase." *Journal of Experimental Botany* 60 (3):815–828. doi: 10.1093/jxb/ern330.

Johnston, W. K., P. J. Unrau, M. S. Lawrence, M. E. Glasner, and D. P. Bartel. 2001. "RNA-catalyzed RNA polymerization: accurate and general RNA-templated primer extension." *Science* 292 (5520):1319–1325. doi: 10.1126/science.1060786.

Joyce, G. F. 1994. Foreword. In *Origins of life: the central concepts*, edited by D. W. Deamer and G. Fleischaker. Boston, USA: Jones and Bartlett. The original reference is a 1992 internal NASA document titled "Exobiology: Discipline Science Plan."

Joyce, G. F. 2002. The antiquity of RNA-based evolution. *Nature* 418:214–221.

Joyce, G. F. 2009. "Evolution in an RNA world." *Cold Spring Harbor Symposia on Quantitative Biology* 74:17–23. doi: 10.1101/sqb.2009.74.004.

Juhas, M., L. Eberl, and J. I. Glass. 2011. "Essence of life: essential genes of minimal genomes." *Trends in Cell Biology* 21 (10):562–568. doi: 10.1016/j.tcb.2011.07.005.

Kaczanowski, S. 2016. "Apoptosis: its origin, history, maintenance and the medical implications for cancer and aging." *Physical Biology* 13 (031001).

Kaczanowski, S., M. Sajid, and S. E. Reece. 2011. "Evolution of apoptosis-like programmed cell death in unicellular protozoan parasites." *Parasites and Vectors* 4:44. doi: 10.1186/1756-3305-4-44.

Kaiser, D. 1986. "Control of multicellular development: *Dictyostelium* and *Myxococcus*." *Annual Review of Genetics* 20:539–566. doi: 10.1146/annurev.ge.20.120186.002543.

Kallius, E. 1931. "Der Zelluntergang als Mechanismus bei der Histio-and Morphogenese." *Verh Anat Ges Suppl Anat Anz* 72:10–22.

Kanavarioti, A., P. A. Monnard, and D. W. Deamer. 2001. "Eutectic phases in ice facilitate nonenzymatic nucleic acid synthesis." *Astrobiology* 1 (3):271–281. doi: 10.1089/15311070152757465.

Kant, I. 1996. *Critique of pure reason*. Unified ed. Translated by W. S. Pluhar. Indianapolis, USA: Hackett Publishing Company. Originally published 1781.

Kapsetaki, S. E., R. M. Fisher, and S. A. West. 2016. "Predation and the formation of multicellular groups in algae." *Evolutionary Ecology Research* 17:651–669.

Kapsetaki, S. E., A. Tep, and S. A. West. 2017. "How do algae form multicellular groups?" *Evolutionary Ecology Research* 18:663–675.

Katsnelson, M. I., Y. I. Wolf, and E. V. Koonin. 2017. "Towards physical principles of biological evolution." *Physica Scripta* 93:043001.

Kazazian, H. H., Jr. 2004. "Mobile elements: drivers of genome evolution." *Science* 303 (5664):1626–1632. doi: 10.1126/science.1089670.

Kempthorne, O. 1983. "Evaluation of current population genetics theory." *American Zoologist* 23:111–121.

Ketcham, R. 2019. "Task allocation and the logic of research questions: how ants challenge human sociobiology." *Biological Theory* 14:52–68.

Khan, N. A., J. Iqbal, and R. Siddiqui. 2015. "Stress management in cyst-forming free-living protists: programmed cell death and/or encystment." *BioMed Research International* 2015:437534. doi: 10.1155/2015/437534.

Kidwell, M. G., and D. R. Lisch. 2000. "Transposable elements and host genome evolution." *Trends in Ecology and Evolution* 15 (3):95–99.

Kiel, J. A. 2010. "Autophagy in unicellular eukaryotes." *Philosophical Transactions of the Royal Society London B* 365 (1541):819–830. doi: 10.1098/rstb.2009.0237.

King, A., and E. Gottlieb. 2009. "Glucose metabolism and programmed cell death: an evolutionary and mechanistic perspective." *Current Opinion in Cell Biology* 21 (6):885–893. doi: 10.1016/j.ceb.2009.09.009.

Kirk, D. L. 2005. "A twelve-step program for evolving multicellularity and a division of labor." *Bioessays* 27 (3):299–310. doi: 10.1002/bies.20197.

Klim, J., A. Gladki, R. Kucharczyk, U. Zielenkiewicz, and S. Kaczanowski. 2018. "Ancestral state reconstruction of the apoptosis machinery in the common ancestor of eukaryotes." *G3 (Bethesda)* 8 (6):2121–2134. doi: 10.1534/g3.118.200295.

Knuuttila, T., and A. Loettgers. 2017. "What are definitions of life good for? Transdisciplinary and other definitions in astrobiology." *Biology and Philosophy* 32:1185–1203.

Kolb, V. M. 2016. "Origins of life: chemical and philosophical approaches." *Evolutionary Biology* 43:506–515.

Kolodkin-Gal, I., R. Hazan, A. Gaathon, S. Carmeli, and H. Engelberg-Kulka. 2007. "A linear pentapeptide is a quorum-sensing factor required for mazEF-mediated cell death in *Escherichia coli*." *Science* 318 (5850):652–655. doi: 10.1126/science.1147248.

Koonin, E. V. 2003. "Comparative genomics, minimal gene-sets and the last universal common ancestor." *Nature Reviews Microbiology* 1 (2):127–136. doi: 10.1038/nrmicro751.

Koonin, E. V. 2007. "The biological big bang model for the major transitions in evolution." *Biology Direct* 2:21. doi: 10.1186/1745-6150-2-21.

Koonin, E. V. 2009. "Darwinian evolution in the light of genomics." *Nucleic Acids Research* 37 (4):1011–1034. doi: 10.1093/nar/gkp089.

Koonin, E. V. 2011. *The logic of chance: the nature and origin of biological evolution*. Upper Saddle River, USA: FT Press.

Koonin, E. V. 2016. "Viruses and mobile elements as drivers of evolutionary transitions." *Philosophical Transactions of the Royal Society London B* 371 (1701). doi: 10.1098/rstb.2015.0442.

Koonin, E. V., and L. Aravind. 2002. "Origin and evolution of eukaryotic apoptosis: the bacterial connection." *Cell Death and Differentiation* 9 (4):394–404. doi: 10.1038/sj/cdd/4400991.

Koonin, E.V., and M. Krupovic. 2019. "Origin of programmed cell death from antiviral defense?" *Proceedings of the National Academy of Science USA* 116:16167–16169. doi: 10.1073/pnas.1910303116.

Koonin, E. V., K. S. Makarova, and Y. I. Wolf. 2017. "Evolutionary genomics of defense systems in archaea and bacteria." *Annual Review of Microbiology* 71:233–261. doi: 10.1146/annurev-micro-090816-093830.

Koonin, E. V., and W. Martin. 2005. "On the origin of genomes and cells within inorganic compartments." *Trends in Genetics* 21 (12):647–654. doi: 10.1016/j.tig.2005.09.006.

Koonin, E. V., Y. I. Wolf, and P. Puigbo. 2009. "The phylogenetic forest and the quest for the elusive tree of life." *Cold Spring Harbor Symposia in Quantitative Biology* 74:205–213. doi: 10.1101/sqb.2009.74.006.

Koonin, E. V., and F. Zhang. 2017. "Coupling immunity and programmed cell suicide in prokaryotes: life-or-death choices." *Bioessays* 39 (1):1–9. doi: 10.1002/bies.201600186.

Kozik, C., E. B. Young, C. D. Sandgren, and J. A. Berges. 2019. Cell death in individual freshwater phytoplankton species: relationships with population dynamics and environmental factors. *European Journal of Phycology* 54:369–379.

Krakauer, D. C., and A. Sasaki. 2002. "Noisy clues to the origin of life." *Proceedings Biological Sciences* 269 (1508):2423–2428. doi: 10.1098/rspb.2002.2127.

Kramer, J., and J. Meunier. 2016. "Kin and multilevel selection in social evolution: a never-ending controversy?" *F1000Res* 5. doi: 10.12688/f1000research.8018.1.

Kroemer, G., W. S. El-Deiry, P. Golstein, M. E. Peter, D. Vaux, P. Vandenabeele, B. Zhivotovsky, M. V. Blagosklonny, W. Malorni, R. A. Knight, M. Piacentini, S. Nagata, and G. Melino. 2005. "Classification of cell death: recommendations of the Nomenclature Committee on Cell Death." *Cell Death and Differentiation* 12 (Suppl 2):1463–1467. doi: 10.1038/sj.cdd.4401724.

Kroemer, G., and J. C. Reed. 2000. "Mitochondrial control of cell death." *Nature Medicine* 6 (5):513–519. doi: 10.1038/74994.

Kumari, S., R. P. Rastogi, K. L. Singh, S. P. Singh, and R. P. Sinha. 2008. "DNA damage: detection strategies." *EXCLI Journal* 7:44–62.

Kundert, P., and G. Shaulsky. 2019. "Cellular allorecognition and its roles in *Dictyostelium* development and social evolution." *International Journal of Developmental Biology* 63:383–393.

Laland, K. N. 2015. "On evolutionary causes and evolutionary processes." *Behavioural Processes* 117:97–104. doi: 10.1016/j.beproc.2014.05.008.

Laland, K. N., T. Uller, M. W. Feldman, K. Sterelny, G. B. Muller, A. Moczek, E. Jablonka, and J. Odling-Smee. 2015. "The extended evolutionary synthesis: its structure, assumptions and predictions." *Proceedings Biological Science* 282 (1813):20151019. doi: 10.1098/rspb.2015.1019.

Lane, N. 2008. "Marine microbiology: origins of death." *Nature* 453:583–585.

Lazcano, A. 2010. "Which way to life?" *Origins of Life and Evolution of Biospheres* 40 (2):161–167. doi: 10.1007/s11084-010-9195-0.

Lehmann, L., L. Keller, S. West, and D. Roze. 2007. "Group selection and kin selection: two concepts but one process." *Proceedings of the National Academy of Science USA* 104 (16):6736–6739. doi: 10.1073/pnas.0700662104.

Leicester, H. M. 1940. "Alexander Mikhailovich Butlerov." *Journal of Chemical Education* 17:203–209.

Levin, S. R., S. Gandon, and S. A. West. 2020. "The social coevolution hypothesis for the origin of enzymatic cooperation." *Nature Ecology and Evolution* 4:132–137. doi: 10.1038/s41559-019-1039-3.

Levraud, J. P., M. Adam, M. F. Luciani, C. de Chastellier, R. L. Blanton, and P. Golstein. 2003. "*Dictyostelium* cell death: early emergence and demise of highly polarized paddle cells." *Journal of Cell Biology* 160 (7):1105–1114. doi: 10.1083/jcb.200212104.

Lewis, K. 2000. "Programmed death in bacteria." *Microbiology and Molecular Biology Reviews* 64 (3):503–514.

Lewontin, R. C. 1970. "The units of selection." *Annual Review of Ecology and Systematics* 1:1–18.

Lidgard, S., and L. K. Nyhart, eds. 2017. *Biological individuality: integrating scientific, philosophical, and historical perspectives.* Chicago, USA. University of Chicago Press.

Lin, H., R. J. Kazlauskas, and M. Travisano. 2017. "Developmental evolution facilitates rapid adaptation." *Scientific Reports* 7 (1):15891. doi: 10.1038/s41598-017-16229-0.

Lincoln, T. A., and G. F. Joyce. 2009. "Self-sustained replication of an RNA enzyme." *Science* 323 (5918):1229–1232. doi: 10.1126/science.1167856.

Lloyd, E. A. 2015. "Adaptationism and the logic of research questions: how to think clearly about evolutionary causes." *Biological Theory* 10:343–362.

Lloyd, E. A., and M. J. Wade. 2019. "Criteria for holobionts from community genetics." *Biological Theory* 14:151–170.

Lockshin, R. A., and C. M. Williams. 1964. "Programmed cell death, II: endocrine potentiation of the breakdown of the intersegmental muscles of silkmoths." *Journal of Insect Physiology* 10:643–649.

Lohrmann, R., and L. E. Orgel. 1971. "Urea-inorganic phosphate mixtures as prebiotic phosphorylating agents." *Science* 171 (3970):490–494.

Ludovico, P., F. Madeo, and M. Silva. 2005. "Yeast programmed cell death: an intricate puzzle." *IUBMB Life* 57 (3):129–135. doi: 10.1080/15216540500090553.

Luper, S. 2009. *The philosophy of death.* Cambridge, UK: Cambridge University Press.

Luper, S. 2016. "Death." In *The Stanford encyclopedia of philosophy*, edited by E. N. Zalta, XX–XX. Stanford, USA: Metaphysics Research Lab, Stanford University.

Luque, V. 2017. "One equation to rule them all: a philosophical analysis of the Price equation." *Biology and Philosophy* 32:97–125.

Lürling, M. 2003. "Phenotypic plasticity in the green algae *Desmodesmus* and *Scenedesmus* with special reference to the induction of defensive morphology." *Annals of Limnology* 32:85–101.

Lürling, M., and E. Van Donk. 1996. "Zooplankton-induced unicell-colony transformation in *Scenedesmus acutus* and its effect on growth of herbivore *Daphnia*." *Oecologia* 108 (3):432–437. doi: 10.1007/BF00333718.

Lynch, M. 2007. *The origins of genome architecture.* Sunderland, USA: Sinauer Associates Inc.

Lyon, B. E., and J. M. Eadie. 2000. "Family matters: kin selection and the evolution of conspecific brood parasitism." *Proceedings of the National Academy of Science USA* 97:12942–12944.

Lyon, P. 2007. "From quorum to cooperation: lessons from bacterial sociality for evolutionary theory." *Studies in History and Philosophy of Biological and Biomedical Sciences* 38:820–833.

Madeo, F., D. Carmona-Gutierrez, J. Ring, S. Buttner, T. Eisenberg, and G. Kroemer. 2009. "Caspase-dependent and caspase-independent cell death pathways in yeast." *Biochemical and Biophysical Research Communications* 382 (2):227–231. doi: 10.1016/j.bbrc.2009.02.117.

Madeo, F., E. Frohlich, and K. U. Frohlich. 1997. "A yeast mutant showing diagnostic markers of early and late apoptosis." *Journal of Cell Biology* 139 (3):729–734. doi: DOI 10.1083/jcb.139.3.729.

Madeo, F., E. Frohlich, M. Ligr, M. Grey, S. J. Sigrist, D. H. Wolf, and K. U. Frohlich.

1999. "Oxygen stress: a regulator of apoptosis in yeast." *Journal of Cell Biology* 145 (4):757–767.

Madeo, F., E. Herker, C. Maldener, S. Wissing, S. Lachelt, M. Herlan, M. Fehr, K. Lauber, S. J. Sigrist, S. Wesselborg, and K. U. Frohlich. 2002. "A caspase-related protease regulates apoptosis in yeast." *Molecular Cell* 9 (4):911–917.

Malik, H. S., W. D. Burke, and T. H. Eickbush. 1999. "The age and evolution of non-LTR retrotransposable elements." *Molecular Biology and Evolution* 16 (6):793–805. doi: 10.1093/oxfordjournals.molbev.a026164.

Mannick, J. B., C. Schonhoff, N. Papeta, P. Ghafourifar, M. Szibor, K. Fang, and B. Gaston. 2001. "S-Nitrosylation of mitochondrial caspases." *Journal of Cell Biology* 154 (6):1111–1116. doi: 10.1083/jcb.200104008.

Manon, S., B. Chaudhuri, and M. Guerin. 1997. "Release of cytochrome c and decrease of cytochrome c oxidase in Bax-expressing yeast cells, and prevention of these effects by coexpression of Bcl-xL." *FEBS Letters* 415 (1):29–32.

Margulis, L. 1981. *Symbiosis in cell evolution*. San Francisco, USA: W. H. Freeman.

Mariscal, C., A. Barahona, N. Aubert-Kato, A. U. Aydinoglu, S. Bartlett, M. L. Cardenas, K. Chandru, C. Cleland, B. T. Cocanougher, N. Comfort, A. Cornish-Bowden, T. Deacon, T. Froese, D. Giovannelli, J. Hernlund, P. Hut, J. Kimura, M. C. Maurel, N. Merino, A. Moreno, M. Nakagawa, J. Pereto, N. Virgo, O. Witkowski, and H. James Cleaves. 2019. "Hidden concepts in the history and philosophy of origins-of-life studies: a workshop report." *Origin of Life and Evolution of the Biosphere* 49 (3):111–145. doi: 10.1007/s11084-019-09580-x.

Mariscal, C., and L. Fleming. 2017. "Why we should care about universal biology." *Biological Theory* 13:121–130.

Marshall, J. A. 2011. "Group selection and kin selection: formally equivalent approaches." *Trends in Ecology and Evolution* 26 (7):325–332. doi: 10.1016/j.tree.2011.04.008.

Maynard Smith, J. 1964. "Group selection and kin selection." *Nature* 201:1145–1147.

Maynard Smith, J., and E. Szathmáry. 1995. *The major transitions in evolution*. Oxford, UK: W. H. Freeman Spektrum.

Mayr, E. 1974. "Teleological and teleonomic, a new analysis." In *A portrait of twenty-five years*, edited by R. S. Cohen and M. W. Wartofsky, 133–159. Boston Studies in the Philosophy of Science. Dordrecht, Netherlands: Springer.

Mayr, E. 1997. "The objects of selection." *Proceedings of the National Academy of Science USA* 94 (6):2091–2094.

McAllister, M. K., and B. D. Roitberg. 1987. "Adaptive suicidal behaviour in pea aphids." *Nature* 328:797. doi: 10.1038/328797b0.

McLennan, D. 2008. "The concept of co-option: why evolution often looks miraculous." *Evolution: Education and Outreach* 1:247–258.

Merchant, S. S., S. E. Prochnik, O. Vallon, E. H. Harris, S. J. Karpowicz, G. B. Witman, A. Terry, A. Salamov, L. K. Fritz-Laylin, L. Marechal-Drouard, W. F. Marshall, L. H. Qu, D. R. Nelson, A. A. Sanderfoot, M. H. Spalding, V. V. Kapitonov, Q. Ren, P. Ferris, E. Lindquist, H. Shapiro, S. M. Lucas, J. Grimwood, J. Schmutz, P. Cardol, H. Cerutti, G. Chanfreau, C. L. Chen, V. Cognat, M. T. Croft, R. Dent, S. Dutcher, E. Fernandez, H. Fukuzawa, D. Gonzalez-Ballester, D. Gonzalez-Halphen, A. Hallmann, M. Hanikenne, M. Hippler, W. Inwood, K. Jabbari, M. Kalanon, R. Kuras, P. A. Lefebvre, S. D. Lemaire, A. V. Lobanov, M. Lohr, A. Manuell, I. Meier, L. Mets, M. Mittag, T. Mittelmeier, J. V. Moroney, J. Moseley, C. Napoli, A. M. Nedelcu, K. Niyogi, S. V. Novoselov, I. T. Paulsen, G. Pazour, S. Purton, J. P. Ral, D. M. Riano-

Pachon, W. Riekhof, L. Rymarquis, M. Schroda, D. Stern, J. Umen, R. Willows, N. Wilson, S. L. Zimmer, J. Allmer, J. Balk, K. Bisova, C. J. Chen, M. Elias, K. Gendler, C. Hauser, M. R. Lamb, H. Ledford, J. C. Long, J. Minagawa, M. D. Page, J. Pan, W. Pootakham, S. Roje, A. Rose, E. Stahlberg, A. M. Terauchi, P. Yang, S. Ball, C. Bowler, C. L. Dieckmann, V. N. Gladyshev, P. Green, R. Jorgensen, S. Mayfield, B. Mueller-Roeber, S. Rajamani, R. T. Sayre, P. Brokstein, I. Dubchak, D. Goodstein, L. Hornick, Y. W. Huang, J. Jhaveri, Y. Luo, D. Martinez, W. C. Ngau, B. Otillar, A. Poliakov, A. Porter, L. Szajkowski, G. Werner, K. Zhou, I. V. Grigoriev, D. S. Rokhsar, and A. R. Grossman. 2007. "The *Chlamydomonas* genome reveals the evolution of key animal and plant functions." *Science* 318 (5848):245–50. doi: 10.1126/science.1143609.

Merlo, L. M., J. W. Pepper, B. J. Reid, and C. C. Maley. 2006. "Cancer as an evolutionary and ecological process." *Nature Reviews Cancer* 6 (12):924–935. doi: 10.1038/nrc2013.

Michod, R. E. 1982. "The theory of kin selection." *Annual Review of Ecology and Systematics* 13:23–55.

Michod, R. E. 1983. "Population biology of the first replicators: on the origin of the genotype, phenotype and organism." *American Zoologist* 23 (1):5–14. doi: 10.1093/icb/23.1.5.

Michod, R. E. 1997. "Evolution of the individual." *American Naturalist* 150 Suppl 1:S5-S21. doi: 10.1086/286047.

Michod, R. E. 1999. *Darwinian dynamics: evolutionary transitions in fitness and individuality*. Princeton, USA: Princeton University Press.

Michod, R. E. 2003. "Cooperation and conflict mediation during the origin of multicellularity." In *Genetic and Cultural Evolution of Cooperation*, edited by P. Hammerstein, 261–307.

Michod, R. E. 2005. "On the transfer of fitness from the cell to the multicellular organism." *Biology and Philosophy* 20 (5):967–987. doi: 10.1007/s10539-005-9018-2.

Michod, R. E. 2006. "The group covariance effect and fitness trade-offs during evolutionary transitions in individuality." *Proceedings of the National Academy of Science USA* 103 (24):9113–9117. doi: 10.1073/pnas.0601080103.

Michod, R. E. 2007. "Evolution of individuality during the transition from unicellular to multicellular life." *Proceedings of the National Academy of Science USA* 104 (Suppl 1):8613–8618. doi: 10.1073/pnas.0701489104.

Michod, R. E., and A. M. Nedelcu. 2003. "On the reorganization of fitness during evolutionary transitions in individuality." *Integrative and Comparative Biology* 43 (1):64–73. doi: 10.1093/icb/43.1.64.

Michod, R. E., and A. M. Nedelcu. 2004. "Cooperation and conflict during the unicellular-multicellular and prokaryotic-eukaryotic transitions." In *Evolution: from molecules to ecosystems*, edited by A. Moya and E. Font, 195–208. Oxford, UK: Oxford University Press.

Michod, R. E., and D. Roze. 1997. "Transitions in individuality." *Proceedings Biological Science* 264 (1383):853–857. doi: 10.1098/rspb.1997.0119.

Miller, S. L. 1953. "A production of amino acids under possible primitive earth conditions." *Science* 117 (3046):528–529.

Miller, S. L., and J. L. Bada. 1988. "Submarine hot springs and the origin of life." *Nature* 334 (6183):609–611. doi: 10.1038/334609a0.

Miller, S. L., and H. C. Urey. 1959a. "Organic compound synthesis on the primitive earth." *Science* 130 (3370):245–251.

Miller, S. L., and H. C. Urey. 1959b. "Origin of life." *Science* 130 (3389):1622–1624. doi: 10.1126/science.130.3389.1622-a.

Minina, E. A., N. S. Coll, H. Tuominen, and P. V. Bozhkov. 2017. "Metacaspases versus caspases in development and cell fate regulation." *Cell Death and Differentiation* 24 (8):1314–1325. doi: 10.1038/cdd.2017.18.

Minina, E. A., J. Staal, V. E. Alvarez, J. A. Berges, I. Berman-Frank, R. Beyaert, K. D. Bidle, F. Bornancin, M. Casanova, J. J. Cazzulo, C. J. Choi, N. S. Coll, V. M. Dixit, M. Dolinar, N. Fasel, C. Funk, P. Gallois, K. Gevaert, E. Gutierrez-Beltran, S. Hailfinger, M. Klemenčič, E. V. Koonin, D. Krappmann, A. Linusson, M. F. M. Machado, F. Madeo, L. A. Megeney, P. N. Moschou, J. C. Mottram, T. Nyström, H. D. Osiewacz, C. M. Overall, K. C. Pandey, J. Ruland, G. S. Salvesen, Y. Shi, A. Smertenko, S. Stael, J. Ståhlberg, M. F. Suárez, M. Thome, H. Tuominen, F. Van Breusegem, R. A. L. van der Hoorn, A. Vardi, B. Zhivotovsky, E. Lam, and P. V. Bozhkov. 2020. "Classification and nomenclature of metacaspases and paracaspases: no more confusion with caspases." *Molecular Cell* 77:927–929.

Mitteldorf, J. 2016. *Aging is a group-selected adaptation: theory, evidence, and medical implications*. London, UK: CRC Press, Taylor and Francis.

Mitteldorf, J., and A. C. Martins. 2014. "Programmed life span in the context of evolvability." *American Naturalist* 184 (3):289–302. doi: 10.1086/677387.

Moharikar, S., J. S. D'Souza, A. B. Kulkarni, and B. J. Rao. 2006. "Apoptotic-like cell death pathway is induced in unicellular chlorophyte *Chlamydomonas reinhardtii* (Chlorophyceae) cells following UV irradiation: detection and functional analyses." *Journal of Phycology* 42:423–433.

Monnard, P. A., and J. W. Szostak. 2008. "Metal-ion catalyzed polymerization in the eutectic phase in water-ice: a possible approach to template-directed RNA polymerization." *Journal of Inorganic Biochemistry* 102 (5–6):1104–1111. doi: 10.1016/j.jinorgbio.2008.01.026.

Morasch, M., J. Liu, C. F. Dirscherl, A. Ianeselli, A. Kuhnlein, K. Le Vay, P. Schwintek, S. Islam, M. K. Corpinot, B. Scheu, D. B. Dingwell, P. Schwille, H. Mutschler, M. W. Powner, C. B. Mast, and D. Braun. 2019. "Heated gas bubbles enrich, crystallize, dry, phosphorylate and encapsulate prebiotic molecules." *Nature Chemistry* 11:779–788. doi: 10.1038/s41557-019-0299-5.

Morasch, M., C. B. Mast, J. K. Langer, P. Schilcher, and D. Braun. 2014. "Dry polymerization of 3′,5′-cyclic GMP to long strands of RNA." *Chembiochem* 15 (6):879–883. doi: 10.1002/cbic.201300773.

Moreno Bergareche, A., and M. Mossio. 2015. *Biological autonomy: a philosophical and theoretical enquiry*. Dordrecht, Netherlands: Springer.

Moritz, E. M., and P. J. Hergenrother. 2007. "Toxin-antitoxin systems are ubiquitous and plasmid-encoded in vancomycin-resistant enterococci." *Proceedings of the National Academy of Science USA* 104 (1):311–316. doi: 10.1073/pnas.0601168104.

Murik, O., A. Elboher, and A. Kaplan. 2014. "Dehydroascorbate: a possible surveillance molecule of oxidative stress and programmed cell death in the green alga *Chlamydomonas reinhardtii*." *New Phytologist* 202 (2):471–484. doi: 10.1111/nph.12649.

Nadell, C. D., J. B. Xavier, and K. R. Foster. 2009. The sociobiology of biofilms. *FEMS Microbiology Reviews* 33:206–224.

Nagel, T. 1974. "What is it like to be a bat?" *Philosophical Review* 83:435–450.

Ndhlovu, A., N. Dhar, N. Garg, T. Xuma, G. C. Pitcher, S. D. Sym, and P. M. Durand 2017. "A red tide forming dinoflagellate *Prorocentrum triestinum*: identification, phylogeny and impacts on St Helena Bay, South Africa." *Phycologia* 56:649–665.

Nedelcu, A. M. 2009. "Comparative genomics of phylogenetically diverse unicellular eukaryotes provide new insights into the genetic basis for the evolution of the programmed cell death machinery." *Journal of Molecular Evolution* 68 (3):256–268.

Nedelcu, A. M., W. W. Driscoll, P. M. Durand, M. D. Herron, and A. Rashidi. 2011. "On the paradigm of altruistic suicide in the unicellular world." *Evolution* 65 (1):3–20. doi: 10.1111/j.1558-5646.2010.01103.x.

Nedelcu, A. M., O. Marcu, and R. E. Michod. 2004. "Sex as a response to oxidative stress: a twofold increase in cellular reactive oxygen species activates sex genes." *Proceedings Biological Science* 271 (1548):1591–1596. doi: 10.1098/rspb.2004.2747.

Nedelcu, A. M., and R. E. Michod. 2003. "Sex as a response to oxidative stress: the effect of antioxidants on sexual induction in a facultatively sexual lineage." *Proceedings Biological Science* 270 (Suppl 2):S136–S139. doi: 10.1098/rsbl.2003.0062.

Nedelcu, A. M., and R. E. Michod. 2011. "Molecular mechanisms of life history tradeoffs and the evolution of multicellular complexity in volvocalean green algae." In *Mechanisms of life history evolution: the genetics and physiology of life history traits and trade-offs*, edited by T. Flatt and A. Heyland, 271–283. Oxford, UK: Oxford University Press.

Niklas, K. J., and S. A. Newman. 2013. "The origins of multicellular organisms." *Evolution and Development* 15 (1):41–52. doi: 10.1111/ede.12013.

Norris, V., and R. Root-Bernstein. 2009. "The eukaryotic cell originated in the integration and redistribution of hyperstructures from communities of prokaryotic cells based on molecular complementarity." *International Journal of Molecular Science* 10 (6):2611–2632. doi: 10.3390/ijms10062611.

Nowak, M. A., A. McAvoy, B. Allen, and E. O. Wilson. 2017. "The general form of Hamilton's rule makes no predictions and cannot be tested empirically." *Proceedings of the National Academy of Science USA* 114:5665–5670.

Nowak, M. A., C. E. Tarnita, and E. O. Wilson. 2010. "The evolution of eusociality." *Nature* 466 (7310):1057–1062. doi: 10.1038/nature09205.

Nuismer, S. 2017. *Introduction to coevolutionary theory*. New York, USA: W. H. Freeman.

Okasha, S. 2006. *Evolution and the levels of selection*. Oxford, UK: Oxford University Press.

Okasha, S. 2016. "The relation between kin and multilevel selection: an approach using causal graphs." *British Journal for the Philosophy of Science* 67:435–470.

Olive, L. S., and C. Stoianovitch. 1975. *The mycetozoans*. New York, USA: Academic.

O'Malley, M., and J. Dupré. 2007. Size doesn't matter: towards a more inclusive philosophy of biology. *Biology and Philosophy* 22:155–191.

Oparin, A. I. 1953. *The origin of life*. New York, USA: Dover.

Orellana, M. V., W. L. Pang, P. M. Durand, K. Whitehead, and N. S. Baliga. 2013. "A role for programmed cell death in the microbial loop." *PLOS ONE* 8:e62595.

Orgel, L. E. 2000. "Self-organizing biochemical cycles." *Proceedings of the National Academy of Science USA* 97 (23):12503–12507. doi: 10.1073/pnas.220406697.

Orgel, L. E. 2004. "Prebiotic chemistry and the origin of the RNA world." *Critical Reviews in Biochemistry and Molecular Biology* 39 (2):99–123. doi: 10.1080/10409230490460765.

Orgel, L. E. 2008. "The implausibility of metabolic cycles on the prebiotic Earth." *PLOS Biology* 6 (1):e18. doi: 10.1371/journal.pbio.0060018.

Oró, J. 1960. "Synthesis of adenine from ammonium cyanide." *Biochemical and Biophysical Research Communications* 2:407–412.

Otsuki, T., D. Uka, H. Ito, G. Ichinose, M. Nii, S. Morita, T. Sakamoto, M. Nishiko, H. Tabunoki, K. Kobayashi, K. Matsuura, K. Iwabuchi, and J. Yoshimura. 2019. "Mass killing by female soldier larvae is adaptive for the killed male larvae in a polyembryonic wasp." *Scientific Reports* 9:7357.

Otto, G. P., M. Y. Wu, N. Kazgan, O. R. Anderson, and R. H. Kessin. 2003. "Macroautophagy is required for multicellular development of the social amoeba *Dictyostelium discoideum*." *Journal of Biological Chemistry* 278 (20):17636–17645. doi: 10.1074/jbc.M212467200.

Palkova, Z. 2004. "Multicellular microorganisms: laboratory versus nature." *EMBO Rep* 5 (5):470–476. doi: 10.1038/sj.embor.7400145.

Pasek, M. A., J. P. Harnmeijer, R. Buick, M. Gull, and Z. Atlas. 2013. "Evidence for reactive reduced phosphorus species in the early Archean ocean." *Proceedings of the National Academy of Science USA* 110 (25):10089–10094. doi: 10.1073/pnas.1303904110.

Pearl, R. 1922. *The biology of death*. Philadelphia, USA: J. B. Lippincott.

Pedersen, K., S. K. Christensen, and K. Gerdes. 2002. "Rapid induction and reversal of a bacteriostatic condition by controlled expression of toxins and antitoxins." *Molecular Microbiology* 45 (2):501–510.

Pence, C. H., and G. Ramsey. 2015. "Is organismic fitness at the basis of evolutionary theory?" *Philosophy of Science* 82:1081–1091.

Penny, D. 2005. "An interpretive review of the origin of life research." *Biology and Philosophy* 20 (4):633–671.

Pentz, J. T., B. P. Taylor, and W. C. Ratcliff. 2016. "Apoptosis in snowflake yeast: novel trait, or side effect of toxic waste?" *Journal of the Royal Society Interface* 13 (118). doi: 10.1098/rsif.2016.0121.

Pepper, J. W., D. E. Shelton, A. Rashidi, and P. M. Durand. 2013. "Are internal death-promoting mechanisms ever adaptive?" *Journal of Phylogenetics and Evolutionary Biology* 1:113.

Pérez Martín, J. M. 2008. *Programmed cell death in protozoa*. Molecular Biology Intelligence Unit Series. New York, USA: Springer-Verlag New York.

Pievani, T. 2014. "Individuals and groups in evolution: Darwinian pluralism and the multilevel selection debate." *Journal of Bioscience* 39 (2):319–325.

Pineda-Krch, M., and K. Lehtilä. 2004. "Costs and benefits of genetic heterogeneity within organisms." *Journal of Evolutionary Biology* 17:1167–1177.

Platinga, A. 1978. *The nature of necessity*. Oxford, UK: Oxford University Press.

Popa, R. 2004. *Between necessity and probability: searching for the definition and origin of life*. Berlin, Germany: Springer-Verlag Berlin Heidelberg.

Powner, M. W., B. Gerland, and J. D. Sutherland. 2009. "Synthesis of activated pyrimidine ribonucleotides in prebiotically plausible conditions." *Nature* 459 (7244):239–242. doi: 10.1038/nature08013.

Price, G. R. 1970. "Selection and covariance." *Nature* 227:520–521.

Price, G. R. 1972. "Extension of covariance selection mathematics." *Annals of Human Genetics* 35:485–490.

Pross, A. 2004. "Causation and the origin of life: metabolism or replication first?" *Origins of Life and Evolution of Biospheres* 34 (3):307–321.

Proto, W. R., G. H. Coombs, and J. C. Mottram. 2013. "Cell death in parasitic protozoa: regulated or incidental?" *Nature Reviews Microbiology* 11 (1):58–66. doi: 10.1038/nrmicro2929.

Prud'homme-Généreux, A. 2013. "What is life? An activity to convey the complexities of this simple question." *American Biology Teacher* 75:53–57.

Queller, D. C. 1992a. "A general model for kin selection." *Evolution* 46 (2):376–380. doi: 10.1111/j.1558-5646.1992.tb02045.x.

Queller, D. C. 1992b. "Quantitative genetics, inclusive fitness, and group selection." *American Naturalist* 139:540–558.

Queller, D. C. 2000. "Relatedness and the fraternal major transitions." *Philosophical Transactions of the Royal Society London B* 355 (1403):1647–1655. doi: 10.1098/rstb.2000.0727.

Queller, D. C. 2016. "Kin selection and its discontents." *Philosophy of Science* 83:861–872.

Radzvilavicius, A. L., and N. W. Blackstone. 2018. "The evolution of individuality revisited." *Biological Reviews of the Cambridge Philosophical Society* 93:1620–1633. doi: 10.1111/brv.12412.

Ramisetty, B. C. M., B. Natarajan, and R. S. Santhosh. 2015. "mazEF-mediated programmed cell death in bacteria: 'what is this?'" *Critical Reviews in Microbiology* 41 (1):89–100. doi: 10.3109/1040841X.2013.804030.

Ramisetty, B. C. M., and R. S. Santhosh. 2017. "Endoribonuclease type II toxin-antitoxin systems: functional or selfish?" *Microbiology* 163 (7):931–939. doi: 10.1099/mic.0.000487.

Ramsey, G. 2016. "The causal structure of evolutionary theory." *Australasian Journal of Philosophy* 94:421–434.

Ramsey, G., and R. Brandon. 2011. "Why reciprocal altruism is not a kind of group selection." *Biology and Philosophy* 26:385–400.

Ramsey, G., and C. H. Pence, eds. 2016. *Chance in evolution*. Chicago, USA: University of Chicago Press.

Ratcliff, W. C., R. F. Denison, M. Borrello, and M. Travisano. 2012. "Experimental evolution of multicellularity." *Proceedings of the National Academy of Science USA* 109 (5):1595–1600.

Ratcliff, W. C., M. D. Herron, K. Howell, J. T. Pentz, F. Rosenzweig, and M. Travisano. 2013. "Experimental evolution of an alternating uni- and multicellular life cycle in *Chlamydomonas reinhardtii*." *Nature Communications* 4.

Ratel, D., S. Boisseau, V. Nasser, F. Berger, and D. Wion. 2001. "Programmed cell death or cell death programme? That is the question." *Journal of Theoretical Biology* 208 (3):385–386. doi: 10.1006/jtbi.2000.2218.

Reece, S. E., L. C. Pollitt, N. Colegrave, and A. Gardner. 2011. "The meaning of death: evolution and ecology of apoptosis in protozoan parasites." *PLOS Pathogens* 7 (12):e1002320. doi: 10.1371/journal.ppat.1002320.

Reeve, H. K., and P. W. Sherman. 1993. "Adaptation and the goals of evolutionary research." *Quarterly Review of Biology* 68:1–32.

Refardt, D., T. Bergmiller, and R. Kümmerli. 2013. "Altruism can evolve when relatedness is low: evidence from bacteria committing suicide upon phage infection." *Proceedings of the Royal Society of London B* 280 (1759):20123035.

Rice, K. C., and K. W. Bayles. 2008. "Molecular control of bacterial death and lysis." *Microbiology and Molecular Biology Reviews* 72:85–109.

Robertson, M. P., and G. F. Joyce. 2012. "The origins of the RNA world." *Cold Spring Harbor Perspectives in Biology* 4:a003608. doi: 10.1101/cshperspect.a003608.

Roff, D. A. 2002. *Life history evolution*. Sunderland, USA: Sinauer Assoc.

Ronai, I., B. P. Oldroyd, and V. Vergoz. 2016. "Queen pheromone regulates programmed cell death in the honey bee worker ovary." *Insect Molecular Biology* 25 (5):646–652. doi: 10.1111/imb.12250.

Rose, M. R., and G. V. Lauder, eds. 1996. *Adaptation*. San Diego, USA: Academic Press.

Rosenberg, E., K. H. Keller, and M. Dworkin. 1977. "Cell density-dependent growth of *Myxococcus xanthus* on casein." *Journal of Bacteriology* 129 (2):770–777.

Rosenbluh, A., and E. Rosenberg. 1989. "Autocide AMI rescues development in dsg mutants of *Myxococcus xanthus*." *Journal of Bacteriology* 171 (3):1513–1518.

Roughgarden, J. 2020. "Holobiont evolution: mathematical model with vertical vs. horizontal microbiome transmission." *Philosophy Theory and Practice in Biology* 12:2.

Roughgarden, J., S. F. Gilbert, E. Rosenberg, I. Zilber-Rosenberg, and E. A. Lloyd. 2017. "Holobionts as units of selection and a model of their population dynamics and evolution." *Biological Theory* 13.

Ruiz-Mirazo, K., and A. Moreno. 2012. "Autonomy in evolution: from minimal to complex life." *Synthese* 185:21–52.

Ruse, M. 1997. "The origin of life: philosophical perspectives." *Journal of Theoretical Biology* 187:473–482. doi: 10.1006/jtbi.1996.0382.

Russell, M. J., and A. J. Hall. 1997. "The emergence of life from iron monosulphide bubbles at a submarine hydrothermal redox and pH front." *Journal of the Geological Society London* 154 (3):377–402.

Saffhill, R. 1970. "Selective phosphorylation of the cis-2′,3′-diol of unprotected ribonucleosides with trimetaphosphate in aqueous solution." *Journal of Organic Chemistry* 35 (9):2881–2883.

Saier, M. H. 2013. "Microcompartments and protein machines in prokaryotes." *Journal of Molecular Microbiology and Biotechnology* 23 (4–5):243–269. doi: 10.1159/000351625.

Sansom, R. 2003. "Constraining the adaptationism debate." *Biology and Philosophy* 18:493–512.

Santelices, B. 1999. "How many kinds of individual are there?" *Trends in Ecology and Evolution* 14 (4):152–155.

Sathe, S., and P. M. Durand. 2016. "Cellular aggregation in *Chlamydomonas* (Chlorophyceae) is chimaeric and depends on traits like cell size and motility." *European Journal of Phycology* 51:129–138.

Sathe, S., S. Kaushik, A. Lalremruata, R. K. Aggarwal, J. C. Cavender, and V. Nanjundiah. 2010. "Genetic heterogeneity in wild isolates of cellular slime mold social groups." *Microbial Ecology* 60 (1):137–148. doi: 10.1007/s00248-010-9635-4.

Sathe, S., M. V. Orellana, N. S. Baliga, and P. M. Durand. 2019. "Temporal and metabolic overlap between lipid accumulation and programmed cell death due to nitrogen starvation in the unicellular chlorophyte *Chlamydomonas reinhardtii*." *Phycological Research* 67:173–183.

Scheiner, S. M., and D. P. Mindell, eds. 2020. *The theory of evolution: principles, concepts, and assumption*. Chicago, USA: University of Chicago Press.

Schwartz, D. A., and D. Lindell. 2017. "Genetic hurdles limit the arms race between *Prochlorococcus* and the T7-like podoviruses infecting them." *ISME Journal* 11 (8):1836–1851. doi: 10.1038/ismej.2017.47.

Segovia, M., L. Haramaty, J. A. Berges, and P. G. Falkowski. 2003. "Cell death in the unicellular chlorophyte *Dunaliella tertiolecta*: a hypothesis on the evolution of apoptosis in higher plants and metazoans." *Plant Physiology* 132 (1):99–105. doi: 10.1104/pp.102.017129.

Segovia, M., M. Teresa Mata, A. Palma, C. García-Gómez, R. Lorenzo, A. Rivera, and F. L. Figueroa. 2015. "*Dunaliella tertiolecta* (Chlorophyta) avoids cell death under ultraviolet radiation by triggering alternative photoprotective mechanisms." *Photochemistry and Photobiology* 91:1389–1140.

Segre, D., D. Ben-Eli, D. W. Deamer, and D. Lancet. 2001. "The lipid world." *Origins of Life and Evolution of Biospheres* 31:119–145.

Seth-Pasricha, M., K. A. Bidle, and K. D. Bidle. 2013. "Specificity of archaeal caspase activity in the extreme halophile *Haloferax volcanii*." *Environmental Microbiology Reports* 5 (2):263–271. doi: 10.1111/1758-2229.12010.

Shakespeare, W. 1603. *The tragicall historie of Hamlet Prince of Denmarke*. London, UK: Printed for N. L. and John Trundell.

Shalini, S., L. Dorstyn, S. Dawar, and S. Kumar. 2015. "Old, new and emerging functions of caspases." *Cell Death and Differentiation* 22 (4):526–539. doi: 10.1038/cdd.2014.216.

Shapiro, R. 2000. "A replicator was not involved in the origin of life." *IUBMB Life* 49 (3):173–176. doi: 10.1080/713803621.

Shapiro, R. 2007. "A simpler origin for life." *Scientific American* 296:46–53.

Silva, R. D., R. Sotoca, B. Johansson, P. Ludovico, F. Sansonetty, M. T. Silva, J. M. Peinado, and M. Corte-Real. 2005. "Hyperosmotic stress induces metacaspase- and mitochondria-dependent apoptosis in *Saccharomyces cerevisiae*." *Molecular Microbiology* 58 (3):824–834. doi: 10.1111/j.1365-2958.2005.04868.x.

Singer, M. A. 2016. "The origins of aging: evidence that aging is an adaptive phenotype." *Current Aging Science* 9 (2):95–115. doi: 10.2174/1874609809666160211124947.

Sinzelle, L., Z. Izsvak, and Z. Ivics. 2009. "Molecular domestication of transposable elements: from detrimental parasites to useful host genes." *Cellular and Molecular Life Sciences* 66 (6):1073–1093. doi: 10.1007/s00018-009-8376-3.

Skillings, D. 2016. "Holobionts and the ecology of organisms: multi-species communities or integrated individuals?" *Biology and Philosophy* 31:875–892.

Skoultchi, A. I., and H. J. Morowitz. 1964. "Information storage and survival of biological systems at temperatures near absolute zero." *Yale Journal of Biology and Medicine* 37:158–163.

Smith, T. J., and S. J. Foster. 1995. "Characterization of the involvement of two compensatory autolysins in mother cell lysis during sporulation of *Bacillus subtilis* 168." *Journal of Bacteriology* 177 (13):3855–3862.

Sober, E. 1993. *The nature of selection*. Chicago, USA: University of Chicago Press.

Sober, E. 2006. *Conceptual issues in evolutionary biology*. 3rd ed. Cambridge, USA: MIT Press.

Sober, E. 2009. "Did Darwin write the *Origin* backwards?" *Proceedings of the National Academy of Science USA* 106 (Suppl 1):10048–10055. doi: 10.1073/pnas.0901109106.

Sober, E., and D. S. Wilson. 1994. "A critical review of philosophical work on the units of selection problem." *Philosophy of Science* 61:534–555.

Sober, E., and D. S. Wilson. 1998. *Unto others: the evolution and psychology of unselfish behavior*. Cambridge, USA: Harvard University Press.

Solari, C. A., V. J. Galzenati, and J. O. Kessler. 2015. "The evolutionary ecology of multicellularity: the volvocine algae as a case study." In *Evolutionary transition to multicellularity: principles and mechanisms*, edited by I. Ruiz-Trillo and A. M. Nedelcu, 201–223. Dordrecht, Netherlands: Springer.

Spungin, D., K. D. Bidle, and I. Berman-Frank. 2019. "Metacaspase involvement in programmed cell death of the marine cyanobacterium *Trichodesmium*." *Environmental Microbiology* 21:667–681.

Stanley, S. M. 1973. "An ecological theory for the sudden origin of multicellular life in the late precambrian." *Proceedings of the National Academy of Science USA* 70 (5):1486–1489.

Stat, M., A. C. Baker, D. G. Bourne, A. M. S. Correa, Z. Forsman, M. J. Huggett, X. Pochon, D. Skillings, R. J. Toonen, M. J. H. van Oppen, and R. D. Gates. 2012. "Molecular delineation of species in the coral holobiont." In *Advances in marine biology*, Vol. 63, edited by M. Lesser, 1–65. Amsterdam, Netherlands: Elsevier.

Stearns, S. C. 1992. *The evolution of life histories*. Oxford, UK: Oxford University Press.

Stigler, S. M. 1980. "Stigler's law of eponymy." *Transactions of the New York Academy of Sciences* 39:147–158.

Strassmann, J. E., R. E. Page, G. E. Robinson, and T. D. Seeley. 2011. "Kin selection and eusociality." *Nature* 471 (7339):E5–E6; author reply E9–E10. doi:10.1038/nature09833.

Strassmann, J. E., and D. C. Queller. 2010. "The social organism: congresses, parties, and committees." *Evolution* 64:605–616.

Sulston, J., R. Lohrmann, L. E. Orgel, and H. T. Miles. 1968. "Nonenzymatic synthesis of oligoadenylates on a polyuridylic acid template." *Proceedings of the National Academy of Science USA* 59 (3):726–733.

Sultan, S. E. 2015. *Organism and environment: ecological development, niche construction, and adaptation*. Oxford, UK: Oxford University Press.

Sun, G., and D. J. Montell. 2017. "Q&A: Cellular near death experiences—what is anastasis?" *BMC Biology* 15:92.

Sundstrom, J. F., A. Vaculova, A. P. Smertenko, E. I. Savenkov, A. Golovko, E. Minina, B. S. Tiwari, S. Rodriguez-Nieto, A. A. Zamyatnin Jr., T. Valineva, J. Saarikettu, M. J. Frilander, M. F. Suarez, A. Zavialov, U. Stahl, P. J. Hussey, O. Silvennoinen, E. Sundberg, B. Zhivotovsky, and P. V. Bozhkov. 2009. "Tudor staphylococcal nuclease is an evolutionarily conserved component of the programmed cell death degradome." *Nature Cell Biology* 11 (11):1347–1354. doi: 10.1038/ncb1979.

Szathmáry, E. 2006. "The origin of replicators and reproducers." *Philosophical Transactions of the Royal Society London B* 361 (1474):1761–1776. doi: 10.1098/rstb.2006.1912.

Szathmáry, E. 2015. "Toward major evolutionary transitions theory 2.0." *Proceedings of the National Academy of Science USA* 112 (33):10104–10111. doi: 10.1073/pnas.1421398112.

Szathmáry, E., and L. Demeter. 1987. "Group selection of early replicators and the origin of life." *Journal of Theoretical Biology* 128 (4):463–486.

Szathmáry, E., and J. Maynard Smith. 1997. "From replicators to reproducers: the first major transitions leading to life." *Journal of Theoretical Biology* 187 (4):555–571. doi: 10.1006/jtbi.1996.0389.

Szostak, J. W. 2012. "Attempts to define life do not help to understand the origin of life." *Journal of Biomolecular Structure and Dynamics* 29 (4):599–600. doi: 10.1080/073911012010524998.

Takeuchi, N., and P. Hogeweg. 2009. "Multilevel selection in models of prebiotic evolution II: a direct comparison of compartmentalization and spatial self-organization." *PLOS Computational Biology* 5 (10):e1000542. doi: 10.1371/journal.pcbi.1000542.

Tatischeff, I., M. Bomsel, C. de Paillerets, H. Durand, B. Geny, D. Segretain, E. Turpin, and A. Alfsen. 1998. "*Dictyostelium discoideum* cells shed vesicles with associated DNA and vital stain Hoechst 33342." *Cellular and Molecular Life Sciences* 54 (5):476–487. doi: 10.1007/s000180050176.

Teresa Mata, M., A. Palma, C. García-Gómez, M. López-Parages, V. Vázquez, I. Cheng-Sánchez, F. Sarabia, F. López-Figueroa, C. Jiménez, and M. Segovia. 2019. "Type II-metacaspases are involved in cell stress but not in cell death in the unicellular green alga *Dunaliella tertiolecta*." *Microbial Cell* 6:494–508.

Teter, S. A., and D. J. Klionsky. 2000. "Transport of proteins to the yeast vacuole: autophagy, cytoplasm-to-vacuole targeting, and role of the vacuole in degradation." *Seminars in Cell and Developmental Biology* 11 (3):173–179. doi: 10.1006/scdb.2000.0163.

Thackeray, J. F., and S. Dykes. 2016. "Morphometric analyses of hominoid crania, probabilities of conspecificity and an approximation of a biological species constant." *HOMO* 67:1–10.

Thackeray, J. F., and C. M. Schrein. 2017. "A probabilistic definition of a species, fuzzy boundaries and 'sigma taxonomy.'" *SA Journal of Science* 113:24–25.

Thamatrakoln, K., B. Bailleul, C. M. Brown, M. Y. Gorbunov, A. B. Kustka, M. Frada, P. A. Joliot, P. G. Falkowski, and K. D. Bidle. 2013. "Death-specific protein in a marine diatom regulates photosynthetic responses to iron and light availability." *Proceedings of the National Academy of Science USA* 110 (50):20123–20128. doi: 10.1073/pnas.1304727110.

Thomas, H., H. J. Ougham, C. Wagstaff, and A. D. Stead. 2003. "Defining senescence and death." *Journal of Experimental Botany* 54 (385):1127–1132.

Thompson, N. S. 2000. "Shifting the natural selection metaphor to the group level." *Behavior and Philosophy* 28 (1–2):83–101.

Toner, J. D., and D. C. Catling. 2020. "A carbonate-rich lake solution to the phosphate problem of the origin of life." *Proceedings of the National Academy of Science USA*. 117 (2): 883–888. doi: 10.1073/pnas.1916109117.

Torgler, C. N., M. de Tiani, T. Raven, J. P. Aubry, R. Brown, and E. Meldrum. 1997. "Expression of bak in *S. pombe* results in a lethality mediated through interaction with the calnexin homologue Cnx1." *Cell Death and Differentiation* 4 (4):263–271. doi: 10.1038/sj.cdd.4400239.

Towe, R. M. 1981. "Environmental conditions surrounding the origin and early archean evolution of life: a hypothesis." *Precambrian Research* 16:1–10.

Toyama, M. 2001. "Adaptive advantages of matriphagy in the foliage spider, *Chiracanthium japonicum* (Araneae: Clubionidae)." *Journal of Ethology* 19:69–74.

Travis, J. M. 2004. "The evolution of programmed death in a spatially structured population." *Journal of Gerontology Series A* 59 (4):301–305.

Tsiatsiani, L., F. Van Breusegem, P. Gallois, A. Zavialov, E. Lam, and P. V. Bozhkov. 2011. "Metacaspases." *Cell Death and Differentiation* 18 (8):1279–1288. doi: 10.1038/cdd.2011.66.

Uren, A. G., K. O'Rourke, L. A. Aravind, M. T. Pisabarro, S. Seshagiri, E. V. Koonin, and V. M. Dixit. 2000. "Identification of paracaspases and metacaspases: two ancient families of caspase-like proteins, one of which plays a key role in MALT lymphoma." *Molecular Cell* 6 (4):961–967.

Vachova, L., F. Devaux, H. Kucerova, M. Ricicova, C. Jacq, and Z. Palkova. 2004. "Sok2p transcription factor is involved in adaptive program relevant for long term survival of *Saccharomyces cerevisiae* colonies." *Journal of Biological Chemistry* 279 (36):37973–37981. doi: 10.1074/jbc.M404594200.

Vachova, L., and Z. Palkova. 2005. "Physiological regulation of yeast cell death in multicellular colonies is triggered by ammonia." *Journal of Cell Biology* 169 (5):711–717. doi: 10.1083/jcb.200410064.

van de Mieroop, M. 2015. *Philosophy before the Greeks: the pursuit of truth in ancient Babylonia*. Princeton, USA: Princeton University Press.

van der Giezen, M. 2011. "Mitochondria and the rise of eukaryotes." *BioScience* 61:594–601.

van der Gulik, P. T., and D. Speijer. 2015. "How amino acids and peptides shaped the RNA world." *Life (Basel)* 5 (1):230–246. doi: 10.3390/life5010230.

Van Dolah, F. M. 2000. "Marine algal toxins: origins, health effects, and their increased occurrence." *Environmental Health Perspectives* 108 (Suppl 1):133–141. doi: 10.1289/ehp.00108s1133.

Van Hautegem, T., A. J. Waters, J. Goodrich, and M. K. Nowack. 2015. "Only in dying, life: programmed cell death during plant development." *Trends in Plant Science* 20 (2):102–113. doi: 10.1016/j.tplants.2014.10.003.

van Holde, K. E. 1980. "The origin of life: a thermodynamic critique." In: *The origin of life and evolution*, edited by H. O. Halvorson and K. E. van Holde, 31–46. New York, USA: Alan R. Liss.

van Valen, L. 2009. "How ubiquitous is adaptation? A critique of the epiphenomenist program." *Biology and Philosophy* 24:267–280.

van Zandbergen, G., A. Bollinger, A. Wenzel, S. Kamhawi, R. Voll, M. Klinger, A. Müller, C. Hölscher, M. Herrmann, and D. Sacks. 2006. "Leishmania disease development depends on the presence of apoptotic promastigotes in the virulent inoculum." *Proceedings of the National Academy of Sciences* 103 (37):13837–13842.

van Zandbergen, G., C. G. Luder, V. Heussler, and M. Duszenko. 2010. "Programmed cell death in unicellular parasites: a prerequisite for sustained infection?" *Trends in Parasitology* 26 (10):477–483. doi: 10.1016/j.pt.2010.06.008.

Vardi, A., D. Eisenstadt, O. Murik, I. Berman-Frank, T. Zohary, A. Levine, and A. Kaplan. 2007. "Synchronization of cell death in a dinoflagellate population is mediated by an excreted thiol protease." *Environmental Microbiology* 9 (2):360–369.

Vardi, A., F. Formiggini, R. Casotti, A. De Martino, F. Ribalet, A. Miralto, and C. Bowler. 2006. "A stress surveillance system based on calcium and nitric oxide in marine diatoms." *PLOS Biology* 4 (3):e60. doi: 10.1371/journal.pbio.0040060.

Vardi, A., L. Haramaty, B. A. Van Mooy, H. F. Fredricks, S. A. Kimmance, A. Larsen, and K. D. Bidle. 2012. "Host-virus dynamics and subcellular controls of cell fate in a natural coccolithophore population." *Proceedings of the National Academy of Science USA* 109 (47):19327–19332. doi: 10.1073/pnas.1208895109.

Vardi, A., B. A. Van Mooy, H. F. Fredricks, K. J. Popendorf, J. E. Ossolinski, L. Haramaty, and K. D. Bidle. 2009. "Viral glycosphingolipids induce lytic infection and cell death in marine phytoplankton." *Science* 326 (5954):861–865. doi: 10.1126/science.1177322.

Vartapetian, A. B., A. I. Tuzhikov, N. V. Chichkova, M. Taliansky, and T. J. Wolpert. 2011. "A plant alternative to animal caspases: subtilisin-like proteases." *Cell Death and Differentiation* 18 (8):1289–1297. doi: 10.1038/cdd.2011.49.

Vavilala, S. L., K. K. Gawde, M. Sinha, and J. S. D'Souza. 2015. "Programmed cell death is induced by hydrogen peroxide but not by excessive ionic stress of sodium chloride in the unicellular green alga *Chlamydomonas reinhardtii*." *European Journal of Phycology* 50:422–438.

Veldhuis, M. J. W., C. P. D. Brussaard, and A. A. M. Noordeloos. 2005. "Living in a *Phaeocystis* colony: a way to be a successful algal species." *Harmful Algae* 4:841–858.

Vesterby, V. 2011. "The intrinsic nature of emergence." *Proceedings of the 55th Annual Meeting of the ISSS*.

Voigt, J., M. Morawski, and J. Wöstemeyer. 2017. "The cytotoxic effects of camptothecin and mastoparan on the unicellular green alga *Chlamydomonas reinhardtii*." *Journal of Eukaryote Microbiology* 64 (6):806–819. doi: 10.1111/jeu.12413.

Voigt, J., and J. Wöstemeyer. 2015. "Protease inhibitors cause necrotic cell death in *Chlamydomonas reinhardtii* by inducing the generation of reactive oxygen species." *Journal of Eukaryote Microbiology* 62 (6):711–721. doi: 10.1111/jeu.12224.

Vostinar, A. E., H. J. Goldsby, and C. Ofria. 2019. "Suicidal selection: programmed cell death can evolve in unicellular organisms due solely to kin selection." *Ecology and Evolution* 9:9129–9136. doi: 10.1002/ece3.5460.

Vreeland, R. H., W. D. Rosenzweig, and D. W. Powers. 2000. "Isolation of a 250 million-year-old halotolerant bacterium from a primary salt crystal." *Nature* 407 (6806):897–900. doi: 10.1038/35038060.

Wachtershauser, G. 1990. "Evolution of the first metabolic cycles." *Proceedings of the National Academy of Science USA* 87:200–204.

Wachtershauser, G. 2003. "From pre-cells to Eukarya—a tale of two lipids." *Molecular Microbiology* 47 (1):13–22.

Wade, M. J. 1977. "An experimental study of group selection." *Evolution* 31 (1):134–153. doi: 10.1111/j.1558-5646.1977.tb00991.x.

Wade, M. J. 1978. "A critical review of the models of group selection." *Quarterly Review of Biology* 53:101–114.

Wade, M. J., and C. J. Goodnight. 1991. "Wright's shifting balance theory: an experimental study." *Science* 253 (5023):1015–1018.

Wagner, G. P., E. M. Erkenbrack, and A. C. Love. 2019. "Stress-induced evolutionary innovation: a mechanism for the origin of cell types." *Bioessays* 41 (4):e1800188. doi: 10.1002/bies.201800188.

Walker, S. I., N. Packard, and G. D. Cody. 2017. "Re-conceptualizing the origins of life." *Philosophical Transactions A Math Phys Eng Sci* 375 (2109). doi: 10.1098/rsta.2016.0337.

Wallace, A. R. 1889. "The action of natural selection in producing old age, decay and death." In *Essays Upon Heredity and Kindred Biological Problems*, edited by A. Weismann. Oxford, UK: Clarendon Press.

Walsh, D. M. 2007. "The pomp of superfluous causes: the interpretation of evolutionary theory." *Philosophy of Science* 74:281–303.

Wang, H., M. Tiezhu, Z. Yu, J. Xiaoli, Q. Lui, and Z. Yu. 2017. "Metacaspases and programmed cell death in *Skeletonema marinoi* in response to silicate limitation." *Journal of Plankton Research* 39:729–743.

Wang, K. J., and J. P. Ferris. 2001. "Effect of inhibitors on the montmorillonite clay-catalyzed formation of RNA: studies on the reaction pathway." *Origins of Life and Evolution of Biospheres* 31 (4–5):381–402.

Watanabe, N., and E. Lam. 2005. "Two *Arabidopsis* metacaspases AtMCP1b and AtMCP2b are arginine/lysine-specific cysteine proteases and activate apoptosis-like cell death in yeast." *Journal of Biological Chemistry* 280 (15):14691–14699. doi: 10.1074/jbc.M413527200.

Wedlich-Soldner, R., and T. Betz. 2018. "Self-organization: the fundament of cell biology." *Philosophical Transaction of the Royal Society London* 373 (1747). doi: 10.1098/rstb.2017.0103.

Weingartner, A., G. Kemmer, F. D. Muller, R. A. Zampieri, M. Gonzaga dos Santos, J. Schiller, and T. G. Pomorski. 2012. "*Leishmania* promastigotes lack phosphatidylserine but bind annexin V upon permeabilization or miltefosine treatment." *PLOS ONE* 7 (8):e42070. doi: 10.1371/journal.pone.0042070.

West, S. A., R. M. Fisher, A. Gardner, and E. T. Kiers. 2015. "Major evolutionary transitions in individuality." *Proceedings of the National Academy of Science USA* 112 (33):10112–10119. doi: 10.1073/pnas.1421402112.

West, S. A., A. S. Griffin, and A. Gardner. 2007. "Social semantics: altruism, cooperation, mutualism, strong reciprocity and group selection." *Journal of Evolutionary Biology* 20 (2):415–432. doi: 10.1111/j.1420-9101.2006.01258.x.

Whittingham, W. F., and K. B. Raper. 1960. "Non-viability of stalk cells in *Dictyostelium*." *Proceedings of the National Academy of Science USA* 46 (5):642–649.

Williams, G. C. 1966. *Adaptation and natural selection*. Princeton, USA: Princeton University Press.

Williams, G. C. 1992. *Natural selection: domains, levels, challenges*. Oxford, UK: Oxford University Press.

Wills, P. R., and C. W. Carter. 2018. "Insuperable problems of the genetic code initially emerging in an RNA world." *Biosystems* 164:155–166.

Wilson, D. S. 1975. "A theory of group selection." *Proceedings of the National Academy of Science USA* 72 (1):143–146.

Wireman, J. W., and M. Dworkin. 1977. "Developmentally induced autolysis during fruiting body formation by *Myxococcus xanthus*." *Journal of Bacteriology* 129 (2):798–802.

Woese, C. R. 1980. "An alternative to the Oparin view of the primeval sequence." In: *The origin of life and evolution*, edited by H. O. Halvorson and K. E. van Holde. New York, USA: Alan R. Liss.

Woese, C. R., and G. E. Fox. 1977. "Phylogenetic structure of the prokaryotic domain: the primary kingdoms." *Proceedings of the National Academy of Science USA* 74 (11):5088–5090.

Wolf, Y. I., M. I. Katsnelson, and E. V. Koonin. 2018. "Physical foundations of biological complexity." *Proceedings of the National Academy of Science USA* 115 (37):E8678–E8687. doi: 10.1073/pnas.1807890115.

Wolkenhauer, O., and J. H. Hofmeyr. 2007. "An abstract cell model that describes the self-organization of cell function in living systems." *Journal of Theoretical Biology* 246 (3):461–476. doi: 10.1016/j.jtbi.2007.01.005.

Wouters, A. 2005. "The function debate in philosophy." *Acta Biotheoretica* 53 (2):123–151. doi: 10.1007/s10441-005-5353-6.

Yamaguchi, Y., J. H. Park, and M. Inouye. 2011. "Toxin-antitoxin systems in bacteria and archaea." *Annual Reviews in Genetics* 45:61–79. doi: 10.1146/annurev-genet-110410-132412.

Yordanova, Z. P., E. J. Woltering, V. M. Kapchina-Toteva, and E. T. Iakimova. 2013. "Mastoparan-induced programmed cell death in the unicellular alga *Chlamydomonas reinhardtii*." *Annals of Botany* 111 (2):191–205. doi: 10.1093/aob/mcs264.

Yorimitsu, T., and D. J. Klionsky. 2005. "Autophagy: molecular machinery for self-eating." *Cell Death and Differentiation* 12 (Suppl 2):1542–1552. doi: 10.1038/sj.cdd.4401765.

Zachar, I., and E. Szathmáry. 2017. "Breath-giving cooperation: critical review of origin of mitochondria hypotheses: major unanswered questions point to the importance of early ecology." *Biology Direct* 12 (1):19. doi: 10.1186/s13062-017-0190-5.

Zheng, W., U. Rasmussen, S. Zheng, X. Bao, B. Chen, Y. Gao, X. Guan, J. Larsson, and B. Bergman. 2013. "Multiple modes of cell death discovered in a prokaryotic (cyanobacterial) endosymbiont." *PLOS ONE* 8 (6):e66147. doi: 10.1371/journal.pone.0066147.

Zuo, Z., Y. Zhu, Y. Bai, and Y. Wang. 2012. "Acetic acid-induced programmed cell death and release of volatile organic compounds in *Chlamydomonas reinhardtii*." *Plant Physiology and Biochemistry* 51:175–184. doi: 10.1016/j.plaphy.2011.11.003.

Zuppini, A., C. Andreoli, and B. Baldan. 2007. "Heat stress: an inducer of programmed cell death in *Chlorella saccharophila*." *Plant Cell Physiology* 48 (7):1000–1009. doi: 10.1093/pcp/pcm070.

Zuppini, A., C. Gerotto, and B. Baldan. 2010. "Programmed cell death and adaptation: two different types of abiotic stress response in a unicellular chlorophyte." *Plant Cell Physiology* 51 (6):884–895. doi: 10.1093/pcp/pcq069.

Zuppini, A., C. Gerotto, R. Moscatiello, E. Bergantino, and B. Baldan. 2009. "*Chlorella saccharophila* cytochrome f and its involvement in the heat shock response." *Journal of Experimental Botany* 60 (14):4189–4200. doi: 10.1093/jxb/erp264.

Index of Names

Page numbers followed by an *f* or a *t* refer to figures and tables, respectively.

Index of Subjects

Page numbers followed by an *f* or a *t* refer to figures and tables, respectively.